Theory and Application of Discrete Element Method to Simulation of Multiphase Flows

Ramesh K. Agarwal · Yali Shao

Theory and Application of Discrete Element Method to Simulation of Multiphase Flows

Springer

Ramesh K. Agarwal
Department of Mechanical Engineering
and Materials Science
Washington University in St. Louis
Saint Louis, MO, USA

Yali Shao
Department of Energy and Mechanical
Engineering
Nanjing Normal University
Nanjing, Jiangsu, China

ISBN 978-3-032-11156-2 ISBN 978-3-032-11157-9 (eBook)
https://doi.org/10.1007/978-3-032-11157-9

This Springer imprint is published by the registered company Springer Nature Switzerland AG
The registered company address is: Gewerbestrasse 11, 6330 Cham, Switzerland

Preface

The Discrete Element Method (DEM) has revolutionized the modeling and understanding of particulate systems, which play a critical role across energy, manufacturing, chemical and pharmaceutical industries. Traditional continuum-based approaches often struggle to capture the complex nature of particle-particle and particle-fluid interactions. By explicitly simulating the motion and contact of individual particles, DEM offers a powerful tool to uncover the underlying physics of granular and multiphase flows. Over the last few decades, the convergence of advanced computational hardware, sophisticated algorithms and the integration of DEM with computational fluid dynamics (CFD) has propelled this technique from a niche research area to a cornerstone of modern engineering design and analysis.

Despite many research papers on DEM and CFD-DEM, there is a need for a comprehensive resource that systematically consolidates the fundamental theoretical framework, the latest advancements in numerical techniques and their diverse industrial applications into a unified volume. This gap increases the difficulty for students and new researchers to quickly grasp the methodology and for experienced practitioners to stay updated on developments beyond their immediate focus. Drawing on the authors' extensive experience in computational particle technology, multiphase flows and high-performance computing, this book provides the state-of-the-art presentation of DEM modeling and simulation. The core of this book is aimed at addressing the scientific and engineering community's need for a thorough overview of methodologies, applications and future directions in the field of DEM for multiphase flows.

This book is the first of its kind to offer such a systematic overview of the field. Its unique feature is the emphasis on seamless progression from core particle-level physics to full-scale industrial application. Key features of the book include:

- Enables readers to build a robust understanding of DEM principles, from particle kinematics and contact models to complex CFD coupling schemes.
- Provides detailed guidance on modeling non-spherical particles, selecting appropriate numerical parameters and implementing large-scale simulations.

- Serves a wide audience, from undergraduate and postgraduate students in chemical, mechanical and process engineering to academic researchers and industry practitioners active in particle technology and multiphase flow research.

This book is structured to provide logical progression of knowledge. Chapters 1 and 2 introduce the history and core principles of the DEM, establishing the theoretical framework for particle kinematics, contact detection and time integration. Chapter 3 details the fundamental contact mechanics models governing particle interactions, including normal, tangential and rolling friction forces. Chapter 4 comprehensively reviews advanced particle shape models, from multi-sphere to polyhedral representations, and their specific contact detection algorithms. Chapters 5 and 6 cover the essential principles of CFD required for modeling the fluid phase in multiphase systems, as well as the critical schemes for coupling DEM and CFD. Chapter 7 expands the framework to include coupled heat and mass transfer mechanisms in the multiphase flow. Chapter 8 presents strategies for large-scale simulation, reviewing high-performance computing hardware, parallelization strategies and numerical acceleration techniques. Subsequent Chaps. 9–17 demonstrate extensive industrial applications, from fluidized beds and aerodynamics systems to pharmaceutical manufacturing and battery technology. Finally, Chap. 18 explores the integration of machine learning and optimization algorithms to enhance DEM model calibration and analysis. In summary, this book provides a thorough foundation in the theory, implementation and application of DEM. It is our sincere hope that this book will become an invaluable resource and a go-to reference, empowering a new generation of engineers and scientists to leverage DEM for innovation and to contribute to the further development of this transformative technology.

The authors wish to express appreciation to the broad research community in the field of particle technology and multiphase flows. The pioneering work and published findings of countless researchers worldwide have formed the essential foundation upon which this book is built. The authors would also like to acknowledge the National Natural Science Foundation of China (Grant No. 52306147) and Natural Science Foundation of Jiangsu Province (Grant No. BK20220377) for its longstanding support of fundamental research in related scientific fields. The authors would also like to express their gratitude to their respective families for their unwavering support that made this arduous undertaking possible. Finally, the authors are grateful to the publishing team at Springer for their professional guidance, patience and expertise throughout the publication process. Their commitment to quality has been evident at every stage.

Saint Louis, USA Ramesh K. Agarwal
Nanjing, China Yali Shao

Contents

Chapter 1
Introduction

1.1 Simulation Methods for Particle Flow

When simulating solid flow, particularly in systems involving granular materials, powders or multiphase flows, two primary computational approaches are widely used: the Eulerian method and the Lagrangian method. The two methods differ fundamentally in how they represent and track the movement of particles in a system.

The Eulerian method treats the flow of solids as a continuous field, similar to how fluids are modeled in traditional computational fluid dynamics (CFD). In this approach, the domain is divided into a fixed grid or mesh, and the properties of the solid material, such as velocity, pressure, and density, are calculated at each point in this grid. Instead of tracking individual particles, the Eulerian method focuses on the overall behavior of the material as it flows through the cells. This method is particularly useful for simulating fluidized beds where the material behaves more like a continuum. The advantage of the Eulerian approach lies in its ability to handle complex boundary conditions and large-scale systems efficiently. However, it can struggle with accurately capturing the detailed dynamics of individual particles, particularly in cases where the flow is sparse or where particle–particle interactions play a crucial role.

In contrast, the Lagrangian method tracks the movement of individual particles or discrete elements in the flow. Each particle is treated as an individual entity with its own position, velocity and other properties, which are updated over time as the particles move and interact with each other and with the surrounding environment. The Lagrangian approach is well-suited for simulating flows where the individual behavior of particles is important, such as in cases of particle mixing, segregation or collisions. Because each particle is tracked separately, the Lagrangian method can provide detailed insights into the microscale dynamics of the flow. However, this method can become computationally expensive, especially when simulating large numbers of particles or when dealing with complex interactions in dense flows.

R. K. Agarwal and Y. Shao, *Theory and Application of Discrete Element Method to Simulation of Multiphase Flows*, https://doi.org/10.1007/978-3-032-11157-9_1

Additionally, managing interactions with fixed boundaries or fluid phases can be more challenging in a Lagrangian framework compared to the Eulerian approach.

1.2 Origins and History of DEM

The DEM employs the Lagrangian framework to model the behavior of systems composed of discrete particles. DEM is widely utilized in fields such as geomechanics, material science, chemical engineering and pharmaceutical manufacturing. The origin and development of DEM are rooted in the need to understand and predict the behavior of complex systems of particles, ranging from soil mechanics to the movement of grains in industrial processes.

The DEM has its roots in the early work of Cundall (1971), who initially developed the method to analyze the mechanical behavior of rock masses. At that time, traditional continuum-based approaches were insufficient for modeling discontinuous systems like fractured rocks or granular materials. Cundall's pioneering work introduced a numerical framework where individual particles or blocks could be treated as distinct entities interacting through contact forces, enabling the simulation of large deformations and discontinuities. His original algorithm, called the Distinct Element Method, laid the foundation for what later became known as DEM.

In the 1980s, DEM gained increasing attraction as researchers recognized its potential for simulating granular materials, which are inherently discontinuous and challenging to model using continuum mechanics. Cundall and Strack [1] further advanced DEM by introducing a particle-based approach for soil and granular flow simulations. Their work introduced key concepts such as contact detection algorithms, force–displacement laws and time-stepping schemes, which remain fundamental to DEM today. The method was initially limited by computational resources, but as computing power improved, DEM became increasingly viable for larger and more complex systems.

Throughout the 1990s and early 2000s, DEM expanded beyond geomechanics into various engineering and scientific disciplines. Researchers began applying DEM to model industrial processes such as powder handling, pharmaceutical tablet compaction and mining operations. Parallel advancements in high-performance computing and graphical processing units allowed for the simulation of millions of particles, making DEM a powerful tool for both academic research and industrial applications. The development of open-source and commercial DEM software further accelerated its adoption, enabling engineers and scientists to explore particle-scale dynamics in ways that were previously impossible.

Today, DEM is a well-established numerical technique with a rich history of interdisciplinary contributions. From its origins in rock mechanics to its widespread use in chemical, civil and mechanical engineering, DEM continues to evolve with new computational techniques, hybrid modeling approaches and machine learning integrations. Its historical development reflects the growing need for high-fidelity simulations of particulate systems, driving innovation across multiple fields.

1.3 Applications and Advancements of DEM

Following the initial work by Cundall and Strack, DEM has evolved into a powerful computational tool with diverse applications across engineering and scientific disciplines. Its ability to model granular and particulate systems at the particle scale has led to significant advancements in both fundamental research and industrial processes.

One of the most prominent applications is in fluidized bed reactors, which are crucial in chemical engineering, energy production and material processing. Conventional fluidized beds, where gas or liquid flows through a granular material to induce fluid-like behavior, are simulated using DEM coupled with CFD in DEM-CFD approaches. This allows researchers to optimize bed hydrodynamics, particle mixing and heat transfer. Additionally, DEM has been applied to magnetized and electrostatic fluidized beds, where external magnetic or electric fields influence particle behavior. DEM simulations help in understanding field-induced particle clustering, segregation and enhanced reaction kinetics. In dense granular systems, DEM plays a crucial role in understanding phenomena such as shear flow. Industries dealing with bulk material handling, such as agriculture and construction, rely on DEM to optimize silo designs and hopper flows. Additionally, DEM has been applied to aerodynamic particle separation systems, such as cyclone, where particle trajectories and separation efficiencies are analyzed to improve industrial processes like mineral processing and waste recycling. Another significant application is in pump systems, particularly those handling slurries or abrasive granular materials. DEM simulations help predict wear patterns, particle-induced erosion and flow blockages, leading to more durable and efficient pump designs. Similarly, in mining engineering, DEM is used to study rock fragmentation and ore processing, which aids in optimizing equipment performance and reducing operational costs. The pharmaceutical industry extensively employs DEM to study powder behavior during milling, blending, granulation, tableting and coating. Understanding particle cohesion, breakage and flow properties is critical for ensuring drug uniformity and manufacturing efficiency. DEM also contributes to battery technology, particularly in optimizing electrode materials composed of granular particles. By modeling the packing, compression and electrochemical behavior of battery powders, DEM helps enhance energy density and durability in lithium-ion and next-generation solid-state batteries.

Beyond these areas, DEM continues to advance through integration with machine learning for faster simulations, multiscale modeling approaches and hybrid DEM-CFD frameworks. Its versatility ensures ongoing relevance in both traditional industries and emerging technologies, making it a cornerstone of modern particle-based simulations.

1.4 Software for DEM

DEM has become a crucial computational tool across various industries and research fields, leading to the development of several software packages tailored to different applications. The commonly used software include EDEM, Particle Flow Code (PFC), Rocky DEM, LAMMPS Improved for General Granular and Granular Heat Transfer Simulations (LIGGGHTS), Yet Another Dynamic Engine (Yade), Multiphase Flow with Interphase eXchanges (MFiX) and Barracuda.

EDEM is a widely used DEM software package developed by Altair, designed for simulating the behavior of granular materials and bulk solids. It is particularly popular in industries such as mining, agriculture and pharmaceuticals, where it is used to optimize processes involving particulate materials, such as material handling, mixing and crushing. EDEM offers a highly intuitive graphical user interface, making it accessible to both beginners and experienced users. The software allows users to set up simulations without requiring extensive programming knowledge. EDEM is optimized for high-performance computing, allowing simulations to run on multi-core processors and graphics processing unit (GPU) clusters. This capability enables users to simulate large-scale systems with millions of particles. EDEM integrates seamlessly with Computer-Aided Design (CAD) software and finite element analysis tools, allowing for coupled simulations. For example, EDEM can be used alongside Altair's HyperMesh and OptiStruct to model the interaction between particles and solid structures. EDEM comes with a comprehensive database of material properties, including those for common bulk materials like coal, sand and gravel. Users can easily customize or add new materials to the database. However, EDEM is a commercial software package, and its licensing fees can be expensive, which might be a barrier for small businesses or academic institutions with limited budgets. Additionally, its high-level interface might limit the ability of advanced users to customize simulations at a deeper level. This can be a disadvantage for researchers who need to implement highly specific or novel models.

PFC, developed by Itasca Consulting Group, is a DEM software package primarily used in geomechanics and rock mechanics. It is particularly well-suited for modeling the behavior of granular materials, jointed rock masses and other systems where the discrete nature of materials plays a critical role. PFC offers a wide range of contact models, including bonded particle models, which are essential for simulating the behavior of materials like concrete, rock and other cohesive materials. The software can model the breakage of particles and bonds, making it highly suitable for fracture mechanics studies. PFC provides a powerful scripting interface through the embedded FISH language, allowing users to customize simulations extensively. This flexibility is beneficial for researchers who need to develop and test new algorithms or models. PFC is also widely recognized for its accuracy in simulating geomechanical problems, making it a popular choice in the mining, civil engineering, and oil and gas industries.

Rocky DEM, developed by Rocky DEM Inc. (now part of Ansys), is a high-performance DEM software specializing in granular and bulk material simulations across industries like mining, pharmaceuticals and energy. Leveraging GPU acceleration, it efficiently simulates millions to billions of particles with complex shapes (spheres, polyhedrons, breakable aggregates) and advanced physics including wear, heat transfer and fluid coupling. Its seamless integration with Ansys Fluent and structural tools enables multiphysics modeling of fluidized beds, crushers and powder processes. Unlike specialized tools like PFC (focused on geomechanics), Rocky DEM offers broad industrial applicability, featuring Python scripting and high-fidelity contact models for realistic particle behavior. Key applications include mineral processing, tablet coating, agricultural grain flow and battery material compaction. With intuitive 3D visualization and rapid computation, Rocky DEM stands out for speed, accuracy and scalability, making it popular choice for engineers optimizing equipment design and granular material handling.

LIGGGHTS is an open-source DEM software package based on the LAMMPS molecular dynamics engine. LIGGGHTS is designed for simulating granular flows and is particularly popular in academic research due to its flexibility and cost-effectiveness. LIGGGHTS is highly scalable and can run on high-performance computing clusters, making it suitable for large-scale simulations with millions of particles. Its performance is comparable to commercial DEM software in terms of speed and efficiency. As an open-source package, LIGGGHTS is highly customizable. Users with programming skills can modify the source code to implement new features, contact models or algorithms. This makes it a powerful tool for research and development. While there is a strong community of users and developers, LIGGGHTS does not come with a graphical user interface, which can make it less user-friendly for beginners.

Yade is an open-source DEM software package primarily developed for research in granular materials and geomechanics. It is known for its flexibility and extensive modeling capabilities, making it a popular choice in academic and research environments. Yade offers a wide range of contact models, including those for cohesive and adhesive materials. The software is particularly strong in simulating geomechanical problems, such as soil-structure interaction, slope stability and rock mechanics. While Yade is flexible and powerful, it may not be as optimized for large-scale simulations as commercial software packages.

MFiX is an open-source software developed by the National Energy Technology Laboratory (NETL) of the US Department of Energy. It is primarily designed for simulating multiphase flows, particularly in fluidized bed. It is widely used in chemical engineering, energy production and environmental applications to model complex processes. One of the main advantages of MFiX is its ability to couple DEM with CFD. This integration allows for detailed simulations of multiphase flows where particle–fluid interactions are critical.

Barracuda Virtual Reactor developed by CPFD Software incorporates DEM within its broader simulation capabilities. It is primarily known for its ability to simulate complex multiphase flow systems, particularly fluidized beds, using a technique called computational particle–fluid dynamics (CPFD). This method combines aspects of both CFD and DEM, making Barracuda a powerful tool for simulating systems involving particles, gases and their interactions. Barracuda's CPFD approach allows for the accurate simulation of systems involving dense particle flows and strong particle–fluid interactions. This is particularly useful for industries that rely on fluidized beds and other multiphase reactors. Additionally, it is capable of simulating chemical reactions in the multiphase flow, making it suitable for applications in chemical engineering and process optimization.

Overall, each of these software packages holds a unique place in the toolkit of engineers and researchers. The best choice depends on the specific requirements of the simulation, the researcher's level of expertise and the resources available.

1.5 Challenges and Future Directions of DEM

Despite its success, DEM faces several challenges that continue to drive research in the field. One major challenge is computational cost, as DEM requires tracking every individual particle and its interactions, leading to high resource demands for large-scale or long-time simulations. While GPU acceleration and parallel computing have improved efficiency, simulations with billions of particles or complex shapes remain computationally expensive. Another challenge is model calibration, as DEM simulations rely heavily on input parameters that are often difficult to measure experimentally. Small errors in these parameters can significantly affect simulation outcomes, requiring advanced calibration technique. Additionally, multiphysics coupling presents difficulties, particularly when integrating DEM with computational fluid dynamics, finite element analysis or thermal models. While DEM-CFD coupling is well-established for many applications such as fluidized beds and pneumatic conveying, challenges persist in accurately resolving particle–fluid interactions at different scales, especially in turbulent or multiphase flows. Similarly, modeling particle breakage, wear and adhesion, which are critical in industries like mining and pharmaceuticals, requires more sophisticated contact models that balance physical realism with computational feasibility.

To overcome these challenges, future advancements in DEM will likely focus on several key areas. High-performance computing and AI-driven optimization will play a crucial role, with developments in GPU computing, reduced-order modeling and neural networks accelerating simulations and improving parameter calibration. Another promising direction is enhanced physics modeling, including better representations of particle shape, breakage and surface chemistry. Furthermore, the integration of DEM with digital twin technology could revolutionize real-time process monitoring and optimization in industries like pharmaceutical.

Reference

1. P.A. Cundall, O.D.L. Strack, A discrete numerical model for granular assemblies. Géotechnique **29**(1), 47–65 (1979). https://doi.org/10.1680/geot.1979.29.1.47

Chapter 2
Theoretical Framework

2.1 Particle Kinematics

Particle kinematics in DEM involves the study of the motion of individual particles, specifically their velocity, acceleration and displacement over time. Each particle in a DEM simulation is tracked individually, with its position updated at each time step based on the forces acting on it.

The velocity of a particle is the rate of change of its position with respect to time. In DEM, the velocity of a particle is updated based on its acceleration, which is derived from the forces acting on it. The velocity vector $\boldsymbol{v}_i$ of particle i at time t is given by:

$$\boldsymbol{v}_i(t + \Delta t) = \boldsymbol{v}_i(t) + \boldsymbol{a}_i(t)\Delta t, \tag{2.1}$$

where $\boldsymbol{a}_i$ is the acceleration of the particle at time t, and Δt is the time step.

Acceleration is the rate of change of velocity with respect to time. It is directly related to the net force acting on the particle through Newton's second law:

$$\boldsymbol{a}_i(t) = \frac{\boldsymbol{F}_i(t)}{m_i}, \tag{2.2}$$

where $\boldsymbol{F}_i(t)$ is the net force acting on particle i at time t, and m_i is the mass of the particle.

In DEM, particles interact through various forces that can be categorized into contact forces, non-contact forces and external forces, as shown in Eq. (2.3). Contact forces arise when particles come into physical contact with each other or with boundaries. The most common contact forces include normal and tangential forces, which are calculated based on the deformation of the particles at the contact point. Non-contact forces act at a distance, without physical contact between particles.

R. K. Agarwal and Y. Shao, *Theory and Application of Discrete Element Method to Simulation of Multiphase Flows*, https://doi.org/10.1007/978-3-032-11157-9_2

$$\boldsymbol{F}_i(t) = \sum \boldsymbol{F}_c + \sum \boldsymbol{F}_{n-c} + \boldsymbol{F}_{\text{ext}}, \tag{2.3}$$

where $\sum \boldsymbol{F}_c$ is the overall particle-particle/wall contact forces acting on particles. $\sum \boldsymbol{F}_{n-c}$ represents non-contact forces between particle-particle/wall. Gravitational force is the most considered non-contact force in granular systems. $\boldsymbol{F}_{\text{ext}}$ represents external forces acting on the particle, which are due to external fields, such as magnetic or electrostatic forces under external magnetic or electric fields.

The angular velocity of a particle is calculated as:

$$\boldsymbol{I}_i \frac{\mathrm{d}\boldsymbol{\omega}_i}{\mathrm{d}t} = \sum \boldsymbol{T}_i, \tag{2.4}$$

where $\boldsymbol{I}_i$ represents the moment of inertia of particle i, $\boldsymbol{\omega}_i$ represents the angular velocity of particle i, and $\boldsymbol{T}_i$ represents torque acting on particle i.

In multiphase flow, the mutual interaction between particle and fluid must be considered to predict the particle kinetics. Compared to the single flow situation, the particle-fluid interaction force $\boldsymbol{F}_{\text{pf}}$ and a torque caused by fluid $\boldsymbol{T}_{\text{pf}}$ are added to the governing equations of the solid phase:

$$m_i \frac{\mathrm{d}\boldsymbol{v}_i}{\mathrm{d}t} = \sum \boldsymbol{F}_c + \sum \boldsymbol{F}_{\text{nc}} + \boldsymbol{F}_{\text{pf}} + \boldsymbol{F}_{\text{ext}} \tag{2.5}$$

$$\boldsymbol{I} \frac{\mathrm{d}\boldsymbol{\omega}_i}{\mathrm{d}t} = \sum \boldsymbol{T}_i + \boldsymbol{T}_{\text{pf}}, \tag{2.6}$$

where the most important particle-fluid interaction force $\boldsymbol{F}_{\text{pf}}$ is the fluid drag force $\boldsymbol{F}_d$. Depending on the specific conditions, other interaction forces, such as unsteady force including Basset force and virtual mass force, lift force including Saffman force and Magnus force and lubrication force may be involved.

2.2 Time Step

The selection of an appropriate time step Δt is a crucial aspect of conducting stable and accurate DEM simulations. The time step determines how frequently the equations of motion for each particle are solved in the simulation. Since DEM employs explicit numerical integration schemes, the choice of Δt directly impacts both the numerical stability and computational efficiency of the simulation. An excessively large time step can lead to numerical instability, while an unnecessarily small time step results in prolonged computational times without significant gains in accuracy [1].

Some key factors influence the choice of the time step, including the stiffness of the particles, particle size, the presence of damping and the nature of the contact model used. Simulations involving hard materials like steel require smaller time steps compared to those with softer materials like rubber, due to the higher stiffness and resulting higher natural frequencies of the particles. Similarly, smaller particles necessitate smaller time steps because their reduced mass leads to faster dynamic responses. Additionally, the inclusion of damping, whether through viscous forces or collision energy dissipation, further restricts the maximum allowable time step to maintain stability.

Traditionally, two primary methods have been used to estimate the critical time step in DEM simulations. The first is the Rayleigh time step criterion, which is derived from the propagation speed of elastic waves through the particles [2]. This approach calculates the time step based on the particle radius, density, shear modulus and Poisson's ratio, providing a simple and widely applicable estimate. However, the Rayleigh criterion does not account for the effects of damping or particle velocity, making it less reliable for highly dynamic or damped systems. The second method, known as the mass-spring criterion, estimates the time step based on the ratio of particle mass to contact stiffness [3]. While this approach works well for linear contact models, it fails to capture the complexities introduced by nonlinear interactions, such as those described by the Hertz-Mindlin contact law, where the contact force varies with the overlap raised to the power of 3/2.

Nonlinear contact models, such as Hertz-Mindlin, present additional challenges for time step selection. The stiffness in such models is not constant but depends on the instantaneous overlap between particles, leading to a dynamic variation in the system's response. Moreover, the presence of damping, whether due to material properties or external factors like liquid bridges, introduces further constraints on the time step. High relative velocities during particle collisions also demand smaller time steps to accurately resolve the rapid changes in force and displacement. To address these challenges, advanced criteria have been developed that explicitly consider the nonlinear nature of the contact forces and the effects of damping. For example, Burns et al. [4] proposed a framework that analyzes the collision process in two phases, that is compression and restitution, and derives the critical time step by ensuring that both phases are adequately resolved. This approach accounts for the initial relative velocity and damping coefficients, providing a more robust estimate for dynamic systems.

2.3 Contact Detection Algorithm

In DEM simulations, contact detection algorithms play a crucial role in determining which particles are in contact and need to be considered for force calculations. Efficient contact detection is critical in large-scale simulations because of the large number of particles and potential contacts. Several types of contact detection algorithms are used, depending on the specific needs of the simulation in terms of accuracy, performance and complexity. Here are the most common types of contact detection algorithms.

2.3.1 Brute-Force Algorithm

The brute-force algorithm is the simplest and most straightforward method for contact detection in DEM. This method involves checking each particle against every other particle in the system to determine if they are in contact. For a system of N particles, this results in $O(N^2)$ comparisons, which can become computationally expensive as the number of particles increases.

Consider two particles i and j, the distance between them is calculated as:

$$d_{ij} = \left|\boldsymbol{r}_i - \boldsymbol{r}_j\right|, \tag{2.7}$$

where $\boldsymbol{r}_i$ and $\boldsymbol{r}_j$ are the position vectors of particles i and j. A contact is detected if d_{ij} is less than the sum of the radii of the two particles $R_i + R_j$. This method ensures accurate contact detection but is highly inefficient for large systems due to its quadratic time complexity.

2.3.2 Neighbor List Algorithm

The neighbor list algorithm, also known as the Verlet list, is an improvement over the brute-force approach. Instead of checking every particle against every other particle, a list of potential neighbors is maintained for each particle. This list contains particles that are in a certain cutoff distance, typically slightly larger than the interaction range ($R_c > R_i + R_j$). The contact detection step then only considers particles in the neighbor list, reducing the number of comparisons. The Verlet list method brings the complexity down to $O(N)$ for building the neighbor list and $O(M)$ for contact detection, where M is the number of neighbors per particle.

2.3.3 Cell-Based Algorithm

The linked-cell algorithm is another widely used method to reduce the computational load of contact detection. In this method, the simulation domain is divided into a grid of cells, with each cell having a size slightly larger than the maximum interaction range between particles. Particles are assigned to cells based on their positions, and only particles in the same or neighboring cells are checked for contact. The cell size l_c is related to the maximum particle radius $R_{\max}$ and the interaction distance d: $l_c \geq 2 \times R_{\max} + d$. For each particle in a cell, contact detection is limited to particles in the same cell or adjacent cells (typically 27 neighboring cells in 3D). The complexity of this algorithm is approximately $O(N)$, making it suitable for large-scale simulations with millions of particles.

2.3.4 *Bounding Volume Hierarchies*

Bounding volume hierarchies (BVH) represent a more sophisticated method for contact detection, particularly useful when dealing with irregular shapes or complex geometries. In BVH, particles or groups of particles are enclosed in bounding volumes such as spheres, boxes or cylinders. These bounding volumes are then organized hierarchically, often in a tree structure. Contact detection starts from the top level of the tree and only refines the search for potential contacts at deeper levels if bounding volumes overlap. The key idea is to quickly eliminate large regions of the simulation domain that do not contain any potential contacts. The computational cost of BVH methods varies depending on the complexity of the particle geometry but can be significantly lower than brute-force methods, particularly for complex shapes.

2.3.5 *Spatial Hashing*

Spatial hashing is an efficient technique for contact detection, particularly in systems with high particle densities. This algorithm maps particles' positions to a discrete grid using a hash function. The grid cells are indexed by hash keys, and particles that share the same hash key or belong to neighboring cells are considered for contact detection. Spatial hashing provides an efficient way to index particles in space, reducing the number of comparisons.

2.3.6 *Multi-scale Contact Detection*

In multiscale contact detection, the simulation domain is processed at multiple resolutions or scales. At the coarsest scale, larger clusters of particles are grouped into bounding volumes, and only those clusters that potentially interact are processed further at finer scales. This approach reduces the number of particles that need to be considered in detail, particularly in scenarios with widely varying particle sizes or where only a subset of particles is involved in interactions.

2.4 Boundary and Initial Conditions

Boundary conditions define how the boundaries of the simulation domain interact with particles. They are crucial in determining whether particles are contained in a region, allowed to move freely, or interact with external forces or constraints. Several types of boundary conditions are commonly used in DEM simulations, depending on the system being modeled:

(1) Periodic boundary conditions

Periodic boundary conditions are commonly used in simulations involving bulk systems or large granular materials where edge effects should be minimized. In periodic boundary conditions, particles that leave one side of the simulation domain re-enter from the opposite side. This creates the illusion of an infinite system and is useful for simulating phenomena such as granular flow in large systems, where the focus is on bulk properties rather than edge effects.

(2) Rigid boundary conditions

Rigid boundary conditions assume that the boundaries of the domain are rigid and that particles cannot pass through them. Instead, when a particle reaches a boundary, it is reflected or stopped, depending on the situation. These boundaries can simulate walls, containers or other physical structures that confine particles.

(3) Absorbing boundary conditions

Absorbing boundaries are used when particles are allowed to leave the simulation domain without returning or interacting with the boundary. This is useful in simulations such as granular flow out of a container where particles escape the system permanently. The particles that reach the boundary are simply removed from the simulation.

(4) Moving boundary conditions

Moving boundary conditions are used in the simulations where boundaries themselves can move or deform. For example, in simulations of mixing or crushing, the boundaries might represent moving machinery such as paddles and rotating drums. The movement of the boundaries interacts with particles, transferring momentum and potentially causing deformation or breakage.

Properly defining initial conditions is also crucial because they significantly affect the final results, especially in transient simulations. Initial conditions in DEM specify the starting configuration of particles in the system, which includes their initial positions, velocities and packing conditions. The initial positions of particles are usually determined based on the system being modeled. For example, in simulations of packing, particles are placed in a predefined arrangement such as a random or regular lattice and then allowed to settle under the influence of gravity. In other cases, particles might be placed in specific regions, such as inside a container or within a defined volume. Initial velocities dictate how particles start moving at the beginning of the simulation. Depending on the application, particles might be initialized with zero velocity or assigned random velocities to simulate thermal agitation or dynamic conditions like flow. For rotational motion, particles may also be assigned initial angular velocities. In many DEM simulations, particularly in packing or compaction studies, the initial packing condition is crucial. Particles can be arranged in loose or dense configurations, often determined by the void fraction or packing fraction.

A dense packing arrangement leads to higher interaction forces early in the simulation, while a loose packing condition may require time for the particles to settle and establish contact.

2.5 Time Integration Schemes

Time integration is a critical component of DEM simulations, governing how particle positions, velocities and forces evolve over discrete time steps. The choice of integration scheme significantly impacts the accuracy, stability and computational efficiency of the simulation. DEM simulations typically employ explicit time integration methods due to their computational efficiency and suitability for modeling dynamic particle interactions. Among these, the Velocity-Verlet algorithm has emerged as the most widely used scheme due to its balance of accuracy, stability and ease of implementation.

The Velocity-Verlet algorithm is a symplectic integrator, meaning it conserves energy well over long simulation time, making it particularly suitable for granular systems. This method updates particle positions and velocities in a two-stage process that first computes new positions based on current velocities and accelerations and then updates velocities using the average of current and newly computed accelerations. The algorithm's time-reversible nature helps maintain numerical stability while providing second-order accuracy in time. In practice, the Velocity-Verlet scheme has become the workhorse of DEM simulations, implemented in major DEM software packages due to its reliable performance across a wide range of granular flow scenarios.

Verlet algorithm, a predecessor to Velocity-Verlet algorithm, offers similar accuracy and better memory efficiency by avoiding explicit velocity storage, but it has become less commonly used in modern DEM implementations. The classical Verlet method calculates new positions using positions from two previous time steps and derives velocities indirectly, which can complicate the implementation of certain contact models. However, its memory efficiency still makes it attractive for extremely large-scale simulations where minimizing memory usage is crucial. Both Verlet and Velocity-Verlet methods share the important property of being symplectic, which helps prevent artificial energy growth or dissipation in long simulations. Simpler explicit methods like the Forward Euler scheme, while easy to implement, are rarely used in DEM codes due to their poor energy conservation properties and tendency to become unstable unless using impractically small time steps. These limitations become particularly apparent in systems with stiff particle interactions or when simulating over long durations.

While explicit methods dominate in DEM due to their simplicity and efficiency, some situations may benefit from implicit integration methods. Implicit methods solve the equations of motion by predicting future states and adjusting them based on system constraints, making them more stable for larger time steps. However, implicit methods are computationally more expensive and involve solving large systems of coupled equations, which limits their use in DEM. They are more often used in finite element methods or when DEM is coupled with other simulation techniques.

2.6 Calculation Procedure

The calculation procedure of DEM begins with initialization, where particle properties such as mass, size, position, velocity and material characteristics are defined. Additionally, contact models governing particle interactions such as Hertz-Mindlin for elastic collisions or linear spring-dashpot models are specified. The computational domain is set up with appropriate boundary conditions and external forces.

Once initialization is complete, the contact detection algorithm identifies pairs of particles that are close enough to interact. For each potential contact, the overlap or distance between particles is calculated to determine whether an actual collision occurs. Following this, force calculation is performed for all detected contacts. The normal and tangential contact forces, which may include elastic, viscous damping and frictional components, are computed based on the predefined contact models. Additional forces, such as cohesion or adhesion, may also be incorporated. The total force acting on each particle is then determined.

If rotational dynamics are considered, torque calculation is carried out to account for tangential forces, rolling resistance and other rotational effects. With all forces and torques known, the equations of motion are solved for each particle using numerical integration techniques like the Velocity-Verlet algorithm. These explicit integration schemes update particle accelerations based on the net force and torque, followed by incremental updates to particle positions and velocities over a small time step. Boundary conditions are enforced to handle interactions with walls or periodic domains.

The simulation progresses by looping through these steps including contact detection, force and torque calculation, motion integration and position/velocity updates for each time increment until the desired simulation duration is reached. Finally, output data, including particle trajectories, velocities and forces, are recorded at specified intervals for post-processing and analysis. This iterative procedure allows DEM to accurately model complex granular dynamics from particle collisions to bulk flow behavior.

References

1. K. Washino, E.L. Chan, K. Miyazaki, T. Tsuji, T. Tanaka, Time step criteria in DEM simulation of wet particles in viscosity dominant systems. Powder Technol. **302**, 100–107 (2016). https://doi.org/10.1016/j.powtec.2016.08.018
2. K. L. Johnson, *Contact Mechanics*, (Cambridge University Press, Cambridge, 1985). https://doi.org/10.1017/CBO9781139171731

3. T. Belytschko, An overview of semidiscretization and time integration procedures, in *Computational Methods for Transient Analysis* (North-Holland, 1983), pp. 1–65
4. S.J. Burns, P.T. Piiroinen, K.J. Hanley, Critical time step for DEM simulations of dynamic systems using a Hertzian contact model. Int. J. Numer. Meth. Eng. **119**(5), 432–451 (2019). https://doi.org/10.1002/nme.6056

Chapter 3
Contact Mechanisms

In discrete element simulations, contact models are broadly categorized into two fundamental approaches: the hard-sphere and soft-sphere models, as shown in Fig. 3.1. The hard-sphere model treats collisions as instantaneous, binary events, resolving them using momentum conservation principles and a coefficient of restitution. This model is computationally efficient for rapid, dilute-phase collisions where multi-particle contacts are rare. In contrast, the soft-sphere model allows for a finite contact duration and explicit deformation by permitting particles to overlap slightly. This overlap is then used to calculate contact forces based on a specified force–displacement law. The key distinction lies in their treatment of time and multiple contacts: the hard-sphere model is event-driven and struggles with enduring contacts, while the soft-sphere model is time-driven and inherently capable of handling multiple, simultaneous contacts, making it the prevalent choice for simulating dense granular systems.

In the soft-sphere framework, the models govern how particles respond to collisions, deformations and external loads. Soft-sphere contact models typically involve three fundamental mechanical laws, as shown in Fig. 3.2. The normal contact force–displacement law governing interactions perpendicular to the contact surface, the tangential contact force–displacement law addressing interactions parallel to the contact surface, and the rolling movement law characterizing resistance to rotational motion between particles [2]. These three distinct yet interconnected components collectively define the complete contact behavior in granular systems. Additionally, advanced models may incorporate effects like cohesion, damping or visco-elasticity to represent more complex material behaviors. The selection of a contact model depends on the material properties and the simulation objectives. For instance, hard, dry particles like grains or minerals may be well-represented by elastic or elasto-plastic models, whereas soft, wet or adhesive materials like wet powders require models that include viscous damping or cohesive forces.

R. K. Agarwal and Y. Shao, *Theory and Application of Discrete Element Method to Simulation of Multiphase Flows*, https://doi.org/10.1007/978-3-032-11157-9_3

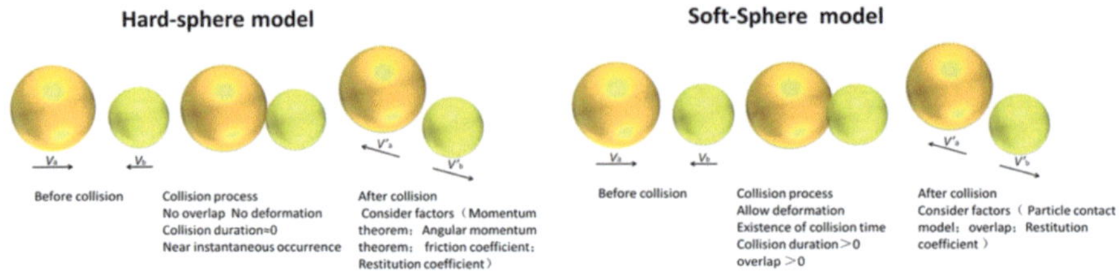

Fig. 3.1 Brief illustration of hard and soft particle model [1]

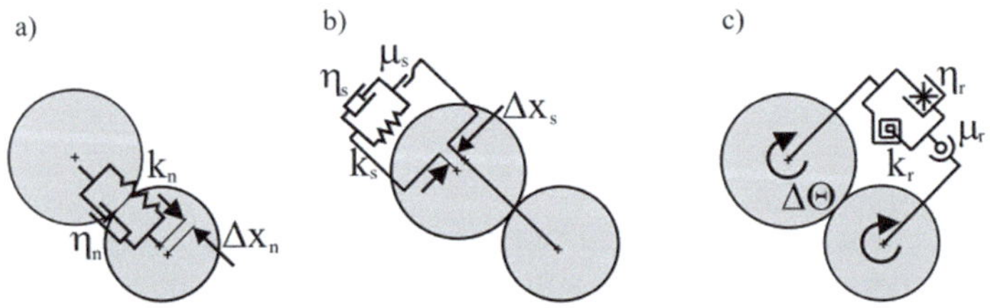

Fig. 3.2 Particle contact models for DEM: **a** normal; **b** tangent; **c** rolling [3]

3.1 Normal Contact Force Modeling

3.1.1 Elastic Contact Model

The elastic contact model describes the interaction between two bodies that deform elastically upon contact. Hertzian contact theory is a classical framework that describes the contact between two elastic bodies, such as spheres or curved surfaces, which undergo small deformations upon contact. Developed by Heinrich Hertz in 1882 [4], this theory assumes that the deformations of the materials remain within their elastic limits, meaning that the bodies return to their original shapes once the contact force is removed. The theory is valid only under specific conditions, such as when the contact area is relatively small compared to the sizes of the objects, and the materials are homogeneous and isotropic.

When two elastic spheres come into contact, Hertzian theory treats the interaction as a non-adhesive contact problem. The contact pressure is distributed elliptically across the contact area, with a maximum at the center and decreasing to zero at the edge, as shown in Fig. 3.3. The surfaces of the spheres deform slightly, creating a small and flat contact region. The radii of curvature of the two bodies, as well as their material properties including Young's modulus and Poisson's ratio, determine the size of this contact region and the magnitude of the contact forces.

Hertzian contact theory provides a relationship between the normal force and the deformation that occurs when the bodies are pressed together. The normal force F_n is related to the normal overlap, which is given by:

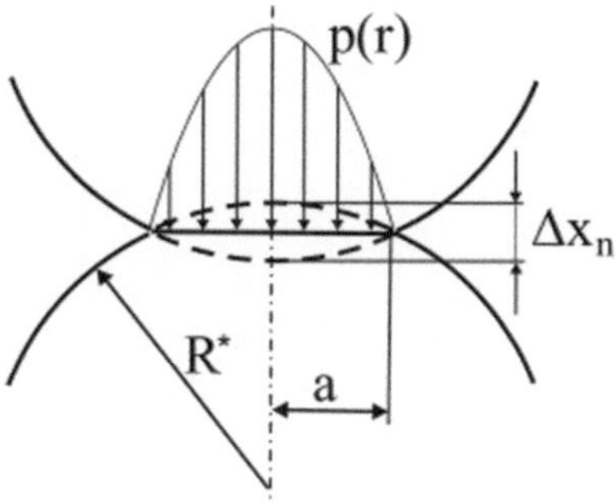

Fig. 3.3 Hertz's contact of two elastic spheres [3]

$$F_n = k_n \Delta x_n^{\frac{3}{2}} = \frac{4}{3} E^* R^{*\frac{1}{2}} \Delta x_n^{\frac{3}{2}}, \tag{3.1}$$

where k_n is the spring stiffness, Δx_n is the normal overlap, E^* is the effective modulus of elasticity and R^* is the effective radius, which are defined as:

$$\frac{1}{E^*} = \frac{1 - \nu_1^2}{E_1} + \frac{1 - \nu_2^2}{E_2} \tag{3.2}$$

$$\frac{1}{R^*} = \frac{1}{R_1} + \frac{1}{R_2}, \tag{3.3}$$

where E_1 and E_2 are the Young's moduli of the two particles. ν_1 and ν_2 are their Poisson's ratios. R_1 and R_2 are the radii.

3.1.2 Elastic–Plastic Contact Model

The elastic–plastic contact model extends the classical Hertzian theory to account for material behavior beyond purely elastic deformation. In real-world applications, contact interactions between bodies often involve pressures that exceed the elastic limit of materials, leading to plastic/permanent deformations. When the contact pressure exceeds the yield strength of the materials, elastic–plastic behavior is observed. The elastic–plastic contact model, therefore, provides a more accurate representation of the contact mechanics for situations where both elastic and plastic deformations coexist.

Figure 3.4 displays the elastic-perfectly plastic contact of spheres according to Thornton and Ning [5]. During the initial elastic stage, the contact force and deformation follow Hertzian theory, where the force is proportional to the overlap raised to the power of 3/2, and the pressure distribution is elliptical. However, once the maximum contact pressure reaches the yield limit p_y, plastic deformation begins. The contact area then divides into two regions: an outer elastic annulus and an inner plastic core where the pressure remains constant at the yield strength. The total

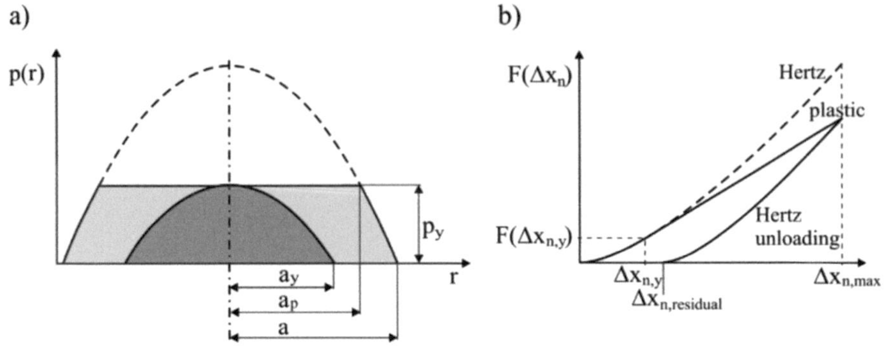

Fig. 3.4 Elastic-perfectly plastic contact of spheres according to Thornton and Ning (1998): **a** pressure distribution, **b** force–displacement relationship [5]

contact force during plastic loading is the sum of the elastic force from the outer annulus and the plastic force from the inner core. Upon unloading, the model assumes that only the elastic portion of the deformation is recovered, following Hertzian mechanics again but with a residual overlap $\Delta x_{n,\text{residual}}$ due to permanent plastic deformation. This residual deformation influences subsequent contacts, though the model does not account for cumulative plastic effects from repeated impacts at the same location. Energy dissipation occurs primarily through plastic work, leading to a velocity-dependent coefficient of restitution, which decreases as impact velocity increases.

The overlap at which yield initiates is given by:

$$\Delta x_{n,y} = \left(\frac{\pi p_y}{1E^*}\right)^2 R^*. \tag{3.4}$$

Consequently, the complete elastic–plastic contact model is defined by the following piecewise function:

$$F_n = \begin{cases} k_n \Delta x_n^{\frac{3}{2}}, & v_n > 0,\ \Delta x_n < \Delta x_{n,y} \text{(Elastic loading)} \\ k_n \Delta x_{n,y}^{\frac{3}{2}} + \pi p_y R^* \left(\Delta x_n - \Delta x_{n,y}\right), & v_n > 0,\ \Delta x_n \geq \Delta x_{n,y} \text{(Plastic loading)} \\ k_n \left(\Delta x_n - \Delta x_{n,\text{residual}}\right)^{\frac{3}{2}}, & v_n < 0,\ \Delta x_n \geq \Delta x_{n,\text{residual}} \text{(Elastic unloading)} \\ 0, & v_n < 0,\ \Delta x_n < \Delta x_{n,\text{residual}} \text{(Separation)}, \end{cases} \tag{3.5}$$

where p_y is the yield pressure, $\Delta x_{n,y}$ is the corresponding overlap of the yield, $\Delta x_{n,\text{residual}}$ is the residual deformation, v_n is the relative velocity in the normal direction at the contact point. More detailed description can be found elsewhere [3].

3.1.3 Visco-Elastic Contact Model

In order to deal with the lack of energy dissipation in the Hertz contact model, the hysteresis damping factor is applied to the normal contact force model, which forms a visco-elastic contact model. The visco-elastic contact model describes the normal contact between two visco-elastic bodies with a spring and dashpot, as shown in Fig. 3.5. In this model, the contact force is divided into two components: an elastic component and a viscous component. This combination allows the model to capture both the energy storage in elastic deformation and the dissipation of energy due to damping, making it particularly useful for simulating dynamic systems where energy loss through inelastic collisions is significant.

(1) Linear spring-dashpot model

A linear model proposed by Kelvin and Voigt, also known as linear spring-dashpot model, is a basic visco-elastic normal contact force model that can account for the energy dissipation of the contact process between two visco-elastic bodies. This model consists of a linear spring element and linear damper element in parallel, and the normal contact force can be expressed as:

$$F_n = k_n \Delta x_n + \eta_n v_{n,} \tag{3.6}$$

where k_n is the normal stiffness coefficient associated with the linear elastic force; η_n is the normal damping coefficient; v_n is the relative velocity in normal direction. The advantage of this model is its simplicity, but its linearity makes it inaccurate for most real contact problems, especially for high-speed impacts. For this linear model, the damping coefficient η_n can be related to the coefficient of restitution e by:

$$\eta_n = -2\ln(e)\sqrt{\frac{k_n m^*}{\pi^2 + \ln^2(e)}}. \tag{3.7}$$

(2) Nonlinear Hertz-based visco-elastic model

On the basis of the Hertz contact model and Kelvin-Voigt linear model, more reasonable nonlinear models have been developed, which can be expressed as:

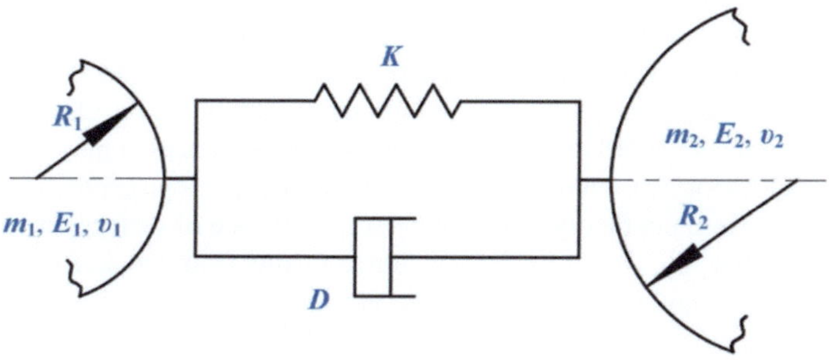

Fig. 3.5 Normal contact model between two visco-elastic bodies [6]

$$F_n = k_n \Delta x_n^{\frac{3}{2}} + c(\Delta x_n)^{\beta} v_{n,} \tag{3.8}$$

where the first term is the Hertzian elastic force, and the second term is a nonlinear viscous dissipative force. c is a damping parameter. β is an exponent that defines the model type.

In the Kuwabara and Kono model ($\beta = 1/2$) [7], the coefficient of restitution e decreases with increasing impact velocity. The damping parameter c is linked to the bulk viscosities B_1 and B_2 of the materials:

$$c = 2\frac{B_1 B_2}{B_1\left(1 - \sigma_2^2\right) + B_2\left(1 - \sigma_1^2\right)}\sqrt{R^*}, \tag{3.9}$$

where B_1, B_2 and σ_1, σ_2 are the bulk viscosities and Poisson's ratios of the two materials, respectively.

In the Tsuji et al. model ($\beta = 1/4$) [8], it assumes a constant coefficient of restitution e independent of impact velocity. The damping parameter c is typically calibrated directly from the measured value of e.

Due to the difficulty in measuring the bulk viscosities, the damping parameter c is often treated as a calibration parameter in DEM simulations to match the observed energy dissipation, even when using the Kuwabara and Kono model.

3.1.4 Adhesion Contact Model

The adhesion contact model is a theoretical framework used to describe the forces involved when two surfaces come into contact, particularly focusing on the adhesive forces that arise due to molecular attraction. These forces can have a significant influence on the mechanics of contact, especially at the micro- and nanoscale. Two commonly used models for adhesion are the Johnson-Kendall-Roberts (JKR) model and the Derjaguin-Muller-Toporov (DMT) model. These models take into account both elastic deformation and adhesive forces between surfaces.

(1) Johnson-Kendall-Roberts model

The JKR model proposed by Johnson, Kendall and Roberts [9] builds on the classical Hertz theory of elastic contact by adding adhesive forces to the problem of two bodies coming into contact. Hertz's theory accounts for the purely elastic deformation of spheres in contact, but it neglects the attractive forces that might arise due to surface energy. The JKR model addresses this limitation by introducing the effects of surface energy, making it particularly useful for situations where adhesion plays a significant role. In the JKR theory, the contact area between two bodies is larger than what would be predicted by Hertzian theory alone, due to the influence of surface energy pulling the bodies together. This additional contact area results from a balance between the elastic deformation and the attractive adhesive forces.

The normal force F_n in the JKR model can be expressed as a function of the contact radius a:

$$F_n = \frac{4}{3}E^*R^{*\frac{1}{2}}a^3 - 3\pi Wa^2, \tag{3.10}$$

where a is the contact radius, which accounts for both elastic and adhesive effects; W is the work of adhesion per unit area, representing the energy required to separate the two surfaces. The first term $\frac{4}{3}E^*R^{*\frac{1}{2}}a^3$ corresponds to the elastic deformation force, similar to Hertzian theory, while the second term $3\pi Wa^2$ accounts for the attractive adhesive forces that act to increase the contact area.

The contact radius a is influenced by both the external load and the adhesive forces. It can be expressed as:

$$a^3 = \frac{R^*}{E^*}\left(F_n + 3\pi WR^* + \sqrt{6\pi WR^*F_n + (3\pi WR^*)^2}\right). \tag{3.11}$$

This equation shows that even when the applied normal force F_n is zero, the contact radius a is still finite due to the adhesive forces, which pull the surfaces together and maintain contact.

The JKR model also predicts a pull-off force, which is the force required to separate the two surfaces. The pull-off force $F_{\text{pull-off}}$ is given by:

$$F_{\text{pull-off}} = -\frac{3}{2}\pi R^*W. \tag{3.12}$$

The JKR model is widely used for describing contact behavior in soft, adhesive materials. It has applications in fields such as polymer science and biological systems. However, it is most applicable when adhesion dominates over elastic deformation, such as in systems with low stiffness or high surface energy. One limitation of the JKR model is that it assumes the adhesive interactions are concentrated near the contact area, which might not be valid for all materials, especially those with long-range interactions.

(2) Derjaguin-Muller-Toporov model

The DMT model [10] is another contact mechanics theory that describes the interaction between two deformable bodies, particularly in the context of adhesion. Like the JKR model, the DMT model is an extension of Hertzian contact theory, but it differs in how it accounts for adhesive forces. While the JKR model assumes that adhesion is concentrated within the contact area, the DMT model assumes that adhesive forces act outside the contact area in the form of long-range surface forces, such as van der Waals forces. The DMT model is typically applied when the contact bodies are stiff or when the adhesive forces are weak compared to the elastic forces, which results in smaller contact areas and more influence from forces outside the contact zone.

The total normal force F_n in the DMT model is the sum of the Hertzian elastic force and the adhesive force acting outside the contact area. This can be written as:

$$F_n = \frac{4}{3} E^* R^{*\frac{1}{2}} a^3 - 2\pi W R^*, \tag{3.13}$$

where the first term $\frac{4}{3} E^* R^{*\frac{1}{2}} a^3$ represents the Hertzian elastic force, and the second term $2\pi WR^*$ is the adhesive force acting outside the contact zone. This adhesive force reduces the overall contact force because it pulls the surfaces together even when there is no external load. a is the contact radius, which is derived from Hertzian theory and does not consider the effects of adhesion within the contact area. In the DMT model, the adhesive forces are assumed not to deform the contact area. Therefore, the pressure distribution within the contact area remains Hertzian, and the relationship between the contact radius a and the elastic force is the same as in Hertz theory. However, the adhesive force outside the contact area effectively changes the load governing the elastic deformation.

Similar to that in JKR model, the pull-off force in the DMT model is given by:

$$F_{\text{pull-off}} = -2\pi R^* W. \tag{3.14}$$

(3) Maugis model

The Maugis model is a comprehensive theory that bridges the gap between the two classical adhesion contact models: the JKR Model and the DMT Model. These two models represent two extremes in the behavior of adhesive contact: JKR for soft materials with strong adhesion and large contact areas, and DMT for hard materials with weak adhesion and small contact areas. The Maugis model accounts for a continuous transition between these two limiting cases and can be applied to a wider range of material properties and conditions.

The Maugis model introduces a dimensionless parameter, known as the Tabor parameter λ, which characterizes the nature of the adhesive contact by comparing the relative contributions of surface forces and elastic deformation. The value of λ indicates the dominant contact regime. If λ is large ($\lambda > 5$), the system behaves more like the JKR model. If λ is small ($\lambda < 0.1$), the system behaves more like the DMT model. For intermediate values of λ, the contact behavior transitions between JKR and DMT, and the Maugis model provides a more accurate description.

Unlike the JKR and DMT models, the Maugis model does not have a simple, closed-form analytical solution for the contact force F_n as a function of contact radius a. Instead, Maugis used a Dugdale approximation to model the adhesive forces, which assumes a constant adhesive stress acting over an annulus just outside the contact area. This formulation leads to a set of parametric equations that must be solved numerically.

The Maugis model also provides a prediction for the pull-off force, which transitions smoothly between the two classical limits. For large λ, the pull-off force corresponds to the JKR pull-off force $F_{\text{pull-off}} = -\frac{3}{2}\pi R^* W$. For small λ,the pull-off force corresponds to the DMT pull-off force $F_{\text{pull-off}} = -2\pi R^* W$.

3.2 Tangential Contact Force Modeling

The tangential contact force model is a fundamental component of DEM that governs the response of particles to sliding, rolling and shearing motions. It is responsible for simulating frictional forces, energy dissipation and the transfer of shear stresses through a granular assembly. The tangential contact behavior between particles is characterized by two distinct regimes: the "stick" regime and the "slip" regime. In the stick regime, which occurs when the tangential force remains below a critical threshold defined by the Coulomb friction limit, the contact point does not experience macroscopic sliding but instead undergoes microscopic elastic shear deformation. This elastic distortion stores energy and allows for the transfer of shear forces across the contact. However, once the applied tangential force exceeds this critical threshold, the contact transitions into the "slip" regime, where the restorative elastic forces are overcome and macroscopic sliding occurs at the contact interface, resulting in energy dissipation through friction.

3.2.1 Linear Spring Coulomb Model

The linear spring coulomb model represents a fundamental and computationally efficient approach to modeling tangential contact forces in DEM simulations. Its core concept simplifies the complex physical interaction into a straightforward mechanical analogy: a linear spring connected in series with a Coulomb friction slider. The spring is responsible for capturing the elastic shear deformation that occurs at the contact point when a tangential force is applied, with its constant stiffness value dictating the relationship between the force and the tangential displacement. This elastic force is, however, bounded by a strict physical limit enforced by the Coulomb slider, which dictates that macroscopic sliding must commence once the magnitude of the tangential force equals the product of the coefficient of static friction and the normal force. The equation is expressed as:

$$\boldsymbol{F}_t = \begin{cases} -k_t \Delta \boldsymbol{x}_t, & |k_t \Delta \boldsymbol{x}_t| < \mu_{p-p} |\boldsymbol{F}_n|, \text{ stick mode} \\ -\mu_{p-p} |\boldsymbol{F}_n| \frac{\Delta \boldsymbol{x}_t}{|\Delta \boldsymbol{x}_t|}, & |k_t \Delta \boldsymbol{x}_t| \geq \mu_{p-p} |\boldsymbol{F}_n|, \text{ slip mode} \end{cases}, \tag{3.15}$$

where k_t is the constant tangential stiffness, μ_{p-p} is the coefficient of static particle–particle friction.

This linear spring coulomb model's primary advantage lies in its simplicity and minimal computational demand, requiring only two parameters, the constant tangential stiffness k_t and the friction coefficient μ. However, this simplicity comes at the cost of physical fidelity, as the assumption of a constant stiffness, independent of the normal load and contact area, is a significant simplification of the true, nonlinear nature of frictional elastic contact.

3.2.2 Mindlin-Deresiewicz Model

The Mindlin-Deresiewicz model [11], also called the elastic-frictional model, is the most physically accurate approach for modeling tangential contact forces between spheres. Its core principle is that tangential stiffness k_t is not constant but increases nonlinearly with normal load due to the expanding contact area, a relationship grounded in Hertzian theory. The nonlinear tangential stiffness k_t is expressed as:

$$k_t = 8G^* \sqrt{R^* \Delta x_{n.}} \tag{3.16}$$

The model's major advancement is its sophisticated handling of history-dependence, meaning it accounts for how simultaneous changes in normal and tangential load affect the contact. It accurately captures micro-slip, which is localized sliding at the contact edges prior to full macroscopic sliding, and how subsequent load changes alter the force–displacement response. This allows it to replicate energy dissipation during loading cycles with high fidelity. However, this physical accuracy requires a computationally expensive incremental solution that tracks the entire contact history, making it best suited for fundamental research rather than large-scale industrial simulations.

3.3 Rolling Friction Modeling

Rolling friction is a crucial mechanism in DEM simulations that accounts for the resistance to rolling motion between particles. Unlike sliding friction, which opposes translational motion at a contact point, rolling friction opposes the relative rotation of particles. In real granular materials, this resistance arises from several sources, including energy dissipation due to micro-slip at the contact interface, plastic deformation of surface asperities, viscous hysteresis in deformable materials, adhesion and the mechanical interlocking of non-spherical shapes. The most common approach to modeling rolling friction in DEM is the constant directional torque model. In this model, a torque is applied to each particle that acts to resist its rolling motion. The magnitude of this torque is proportional to the normal force at the contact and a characteristic length scale, while its direction opposes the relative rolling velocity. For more advanced applications, particularly those involving non-spherical particles or requiring a more physical representation of contact mechanics, alternative models are employed. The virtual spring model introduces a spring at the contact point that generates a resisting torque proportional to the relative rotation between particles, effectively modeling the elastic resistance to rolling. Another sophisticated approach is the contact-independent model, which applies a constant resisting torque directly to a particle's rotational degree of freedom, independent of specific contacts.

3.4 Contact Parameters

3.4.1 Modulus of Elasticity and Poisson's Ratio

Modulus of elasticity also known as Young's modulus E and Poisson's ratio ν are critical parameters used to model the mechanical behavior of individual particles and their interactions. These properties influence how particles deform and transmit forces when they come into contact with one another, which is essential for simulating realistic material behavior in granular systems.

The Young's modulus describes the stiffness of a material and represents the relationship between stress and strain in the linear elastic region of a material. In DEM, this parameter controls how much a particle deforms when subjected to a normal contact force. The higher the Young's modulus, the less the particle will deform under a given force, meaning the particle is stiffer. The expression of E is given as:

$$E = \frac{\sigma}{\varepsilon} = \frac{F/A}{\Delta L/L_0}, \tag{3.17}$$

where σ is the stress; ε is the strain; F is the applied force; A is the cross-sectional are; ΔL is the length change; L_0 is the original length.

In DEM simulations, the contact forces between particles are often calculated using contact models such as the Hertzian contact theory, where Young's modulus plays a key role in determining the normal contact force between two particles. The normal force F_n is typically related to the relative displacement between particles according to a contact law that incorporates the effective Young's modulus of the interacting particles, as given in Eq. (3.1). For two particles with moduli E_1 and E_2, the effective modulus of elasticity E^* is given by Eq. (3.2).

Poisson's ratio ν describes how a material deforms laterally when it is compressed or stretched. It is defined as the ratio of lateral strain to axial strain. In DEM, when two particles come into contact and experience deformation, Poisson's ratio affects the transfer of stress in the tangential direction as the particle deforms under the applied normal load. It is particularly important when dealing with non-isotropic deformations, such as in materials that exhibit different elastic properties in different directions or under different loading conditions.

$$\nu = -\frac{\varepsilon_t}{\varepsilon_l}, \tag{3.18}$$

where ε_t and ε_l are the lateral strain and axial strain, respectively.

3.4.2 Damping Parameters and Restitution Coefficient

Damping represents energy dissipation mechanisms that occur during particle interactions. The most common approach uses viscous damping, where the damping force is proportional to the relative velocity between colliding particles. This is typically implemented through a spring-dashpot contact model, where the spring represents elastic deformation and the dashpot accounts for energy loss. The damping coefficient determines how quickly kinetic energy is dissipated during collisions and directly influences the system's dynamic response. In DEM implementations, damping can be applied in both the normal and tangential directions of contact, with separate coefficients for each direction to capture different energy dissipation mechanisms.

The restitution coefficient quantifies the elasticity of collisions between particles, defined as the ratio of relative velocities after and before impact. This parameter ranges from 0 (perfectly plastic collision) to 1 (perfectly elastic collision). In DEM, the restitution coefficient is often related to the damping coefficient through the contact model parameters. For the commonly used linear spring-dashpot model, the restitution coefficient can be expressed as a function of the damping ratio, demonstrating how energy loss during collisions affects the rebound behavior of particles.

Implementation in DEM simulations involves careful consideration of how these parameters interact. The damping coefficient controls the rate of energy dissipation during continuous contact, while the restitution coefficient governs the energy retained during discrete collision events. These parameters are particularly important in simulations involving dynamic processes such as mixing, segregation or impact, where energy dissipation mechanisms significantly influence the system behavior. The choice of values for these parameters should be based on experimental measurements or careful calibration to ensure the simulation accurately represents the real physical system being modeled. Practical considerations for these parameters include their dependence on material properties, impact velocity and environmental conditions. In many DEM simulations, constant values are used for simplicity, though more advanced models may implement velocity-dependent or material-dependent coefficients to better capture real-world behavior.

3.4.3 Particle–Particle/Particle–Wall Coefficient of Friction

The particle–particle coefficient of friction μ_{p-p} and particle–wall coefficient of friction μ_{p-w} are fundamental parameters in DEM simulations, governing the shear resistance, flow behavior and energy dissipation in granular systems. These coefficients quantify the frictional interactions between contacting bodies, either between individual particles or between particles and boundary surfaces.

The particle–particle coefficient of friction μ_{p-p} defines the resistance to sliding between two particles in contact. In DEM, μ_{p-p} is typically modeled using Coulomb's friction law, where sliding initiates once the tangential force exceeds the product $\mu_{p-p} * F_n$. However, real-world particle interactions are more complex due to factors such as surface roughness, material hardness, moisture and adhesion. For example, rougher surfaces increase μ_{p-p} due to interlocking of asperities; softer materials exhibit higher μ_{p-p} due to greater deformation at contact points; in wet or cohesive materials, liquid bridges or van der Waals forces enhance frictional resistance, requiring models that combine Coulomb friction with adhesion terms. Experimental measurements of μ_{p-p} often involve sliding tests between individual particles or bulk shear tests. However, DEM simulations frequently use calibrated values higher than single-particle measurements to account for additional bulk effects like particle rearrangement and geometric interlocking that are not captured in a single-contact measurement.

The particle–wall coefficient of friction μ_{p-w} describes the frictional resistance between particles and boundary surfaces, such as container walls, chutes or industrial equipment. Unlike μ_{p-p}, μ_{p-w} depends heavily on the wall material's properties, including the surface texture, material pairing and dynamic effects. For example, sand particles sliding on concrete exhibit higher μ_{p-w} than on stainless steel due to differences in hardness and surface energy; at higher velocities, μ_{p-w} may decrease due to reduced contact time or increase due to inertial effects, requiting velocity-dependent friction models. In DEM, μ_{p-w} significantly influences system behavior including flow patterns, segregation, wear and abrasion.

3.4.4 Coefficient of Rolling Friction

The coefficient of rolling friction μ_r accounts for the resistance to rolling motion between particles or between particles and boundaries. Unlike sliding friction, which opposes translational motion, rolling friction resists rotational movement. This phenomenon arises from various microscopic mechanisms, including elastic hysteresis, plastic deformation, surface adhesion and geometric interlocking at contact points. In DEM, rolling friction is typically modeled as an additional torque that opposes the relative rotation between contacting bodies. The value of μ_r depends heavily on material properties. The importance of rolling friction becomes particularly evident in systems dominated by rotational dynamics. In industrial applications like ball mills or rotary kilns, proper calibration of μ_r is essential for predicting power consumption and mixing efficiency. In geotechnical engineering, it influences the stability of granular slopes and the formation of shear bands. For non-spherical particles, rolling friction also helps approximate shape effects. Elongated or angular particles naturally exhibit higher effective rolling resistance due to geometric interlocking.

References

1. Y. Yang et al., Research on vibrating screen screening technology and method based on DEM: a review, in *Springer Science and Business Media Deutschland GmbH* (2024). https://doi.org/10.1007/s42461-024-01080-5
2. H.P. Zhu, Z.Y. Zhou, R.Y. Yang, A.B. Yu, Discrete particle simulation of particulate systems: theoretical developments. Chem. Eng. Sci. **62**(13), 3378–3396 (2007). https://doi.org/10.1016/j.ces.2006.12.089
3. J. Horabik, M. Molenda, in *Parameters and Contact Models for DEM Simulations of Agricultural Granular Materials: A Review* (Academic Press, 2016). https://doi.org/10.1016/j.biosystemseng.2016.02.017
4. H. Hertz, Ueber die Berührung fester elastischer Körper. J. Reine Angew. Math. **1882**(92), 156–171 (1882). https://doi.org/10.1515/crll.1882.92.156
5. C. Thornton, Z. Ning, A theoretical model for the stick/bounce behaviour of adhesive, elastic-plastic spheres. Powder Technol. **99**(2), 154–162 (1998). https://doi.org/10.1016/S0032-5910(98)00099-0
6. S. Ding, B. Jian, Y. Zhang, R. Xia, G. Hu, A normal contact force model for viscoelastic bodies and its finite element modeling verification. Mech. Mach. Theory **181**, 105202 (2023). https://doi.org/10.1016/j.mechmachtheory.2022.105202
7. G. Kuwabara, K. Kono, Restitution coefficient in a collision between two spheres. Jpn. J. Appl. Phys. **26**(8), 1230–1233 (1987) [Online]. Available: https://www.scopus.com/inward/record.uri?eid=2-s2.0-0023393089&partnerID=40&md5=50ba40166470f85b5ce5ca4353267139
8. Y. Tsuji, T. Tanaka, T. Ishida, Lagrangian numerical simulation of plug flow of cohesionless particles in a horizontal pipe. Powder Technol. **71**(3), 239–250 (1992). https://doi.org/10.1016/0032-5910(92)88030-L
9. K.L. Johnson, K. Kendall, A.D. Roberts, Surface energy and the contact of elastic solids. Proc. Roy. Soc. London Series A Math. Physic. Sci. **324**(1558), 301–313 (1971)
10. B.V. Derjaguin, V.M. Muller, Yu.P. Toporov, Effect of contact deformations on the adhesion of particles. J. Colloid Interface Sci. **53**(2), 314–326 (1975). https://doi.org/10.1016/0021-9797(75)90018-1
11. R.D. Mindlin, H. Deresiewicz, Elastic spheres in contact under varying oblique forces. J. Appl. Mech. Trans. ASME 327–344 (1953). https://doi.org/10.1115/1.4010702

Chapter 4
Particle Shape Models and Contact Detection Algorithms

4.1 Sphere-Based Model

Sphere-based models represent a fundamental approach in DEM for approximating non-spherical particles through the use of spherical elements. These models have become widely adopted in particle dynamics simulations due to their computational efficiency in contact detection and force calculation, while providing reasonable approximations of complex particle shapes. The basic principle involves using either multiple overlapping spheres or modified single spheres to mimic the geometric and dynamic characteristics of non-spherical particles. This approach offers a practical compromise between simulation accuracy and computational cost, making it particularly suitable for large-scale industrial applications where simulating millions of particles is necessary. Three main variants of sphere-based models have been developed to address different simulation requirements: the multisphere model, bonded-sphere model and single sphere with rolling resistance model, each offering distinct advantages for specific applications.

4.1.1 Multisphere Model

The multisphere model, alternatively referred to as the clumped sphere approach, represents a fundamental technique in DEM simulations for approximating non-spherical particle geometries. As shown in Fig. 4.1, this method constructs composite particles by rigidly connecting multiple spherical elements, creating assemblies that collectively mimic the shape and mechanical behavior of target non-spherical particles while maintaining the computational advantages of spherical contact mechanics.

The multisphere approach preserves the individual geometric and mechanical properties of constituent spheres while enforcing rigid-body kinematics across the

R. K. Agarwal and Y. Shao, *Theory and Application of Discrete Element Method to Simulation of Multiphase Flows*, https://doi.org/10.1007/978-3-032-11157-9_4

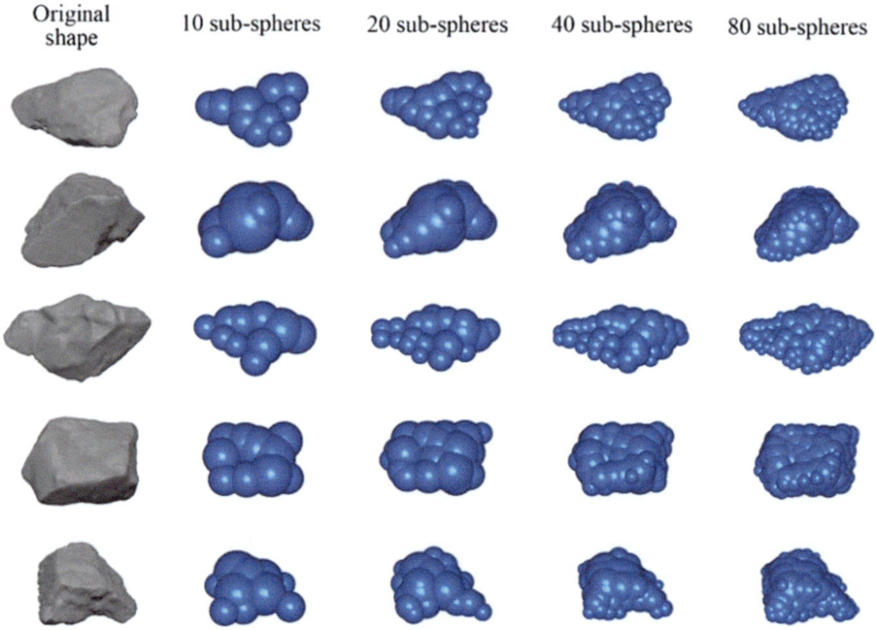

Fig. 4.1 Particles in different shapes described by multisphere model [1]

entire assembly. Each spherical element contributes to the composite particle's mass distribution, inertia properties and contact characteristics. The model calculates the assembly's center of mass and moment of inertia through systematic integration of the component spheres' physical attributes. During simulation, the entire assembly translates and rotates as a singular rigid entity, with fixed relative positions among constituent spheres, which is a simplification that dramatically reduces computational complexity compared to modeling independent spherical elements.

The construction process involves strategically positioning spheres to achieve optimal shape representation. Elongated particles, for instance, are modeled through linear sphere arrangements along major axes, while cubic geometries employ three-dimensional lattice configurations. Controlled sphere overlap creates more continuous surfaces, though excessive overlap requires careful consideration to maintain computational efficiency. The approximation fidelity depends critically on two factors: the number of component spheres and their spatial distribution. Complex geometries with sharp edges or flat facets demand higher sphere counts for accurate representation, whereas smoother contours can be adequately captured with fewer elements. This creates an inherent trade-off between geometric accuracy and computational load that must be carefully balanced for each application.

The model's widespread adoption stems from its direct utilization of established spherical contact algorithms. Contact detection between assemblies occurs through pairwise sphere intersection checks, with force calculations employing conventional spring-dashpot formulations. Normal contact forces derive from sphere overlaps and assigned material properties, while tangential interactions incorporate frictional and damping components. The system aggregates these individual contact forces and torques at the assembly's center of mass, enabling efficient rigid-body dynamics computation.

This modeling approach finds particular utility in industrial simulations involving granular materials, powders and aggregates where particle morphology significantly influences system behavior. Its balanced compromise between shape fidelity and computational efficiency makes it preferable to pure spherical representations in mining, pharmaceutical and agricultural applications. The method's flexibility accommodates diverse particle geometries including elongated, angular and irregular forms that commonly occur in industrial processes.

Several important constraints govern the model's effective application: (1) The inherent approximation nature means shape accuracy directly correlates with sphere count, creating potential accuracy-computation trade-offs; (2) while excelling at external shape representation, the method cannot naturally capture internal material heterogeneity or void structures without specialized modifications; (3) the rigid-body assumption precludes direct simulation of particle deformation or fracture mechanisms; (4) surface roughness artifacts from discrete sphere arrangements may influence particulate flow dynamics in certain scenarios.

4.1.2 Bonded-Sphere Model

The bonded-sphere model represents an advanced extension of the multisphere approach in DEM simulations, designed to overcome the rigid-body limitations of conventional clumped sphere methods. This sophisticated modeling technique connects spherical elements through mechanical bonds that can transmit forces and moments while permitting controlled deformation and eventual breakage under specified conditions, as shown in Fig. 4.2. Unlike the rigid multisphere assemblies, bonded-sphere particles maintain their structural integrity through these dynamic connections until applied stresses exceed predefined thresholds, making the approach particularly valuable for simulating materials that exhibit elastic, plastic or brittle mechanical responses.

The model demonstrates exceptional versatility in capturing complex particulate behaviors across multiple industries. In pharmaceutical manufacturing, it accurately simulates tablet compaction processes and subsequent mechanical integrity, where the bonds represent cohesive forces between powder granules. For agricultural and food processing applications, the method effectively models the breakage patterns of brittle agglomerates during handling and transportation. The bonded-sphere approach has proven particularly valuable in geotechnical engineering, where it reproduces the fracture mechanics of rock masses and the shear behavior of cohesive soils with remarkable fidelity.

Implementation of the bonded-sphere model introduces several computational considerations that distinguish it from rigid multisphere approaches. The additional

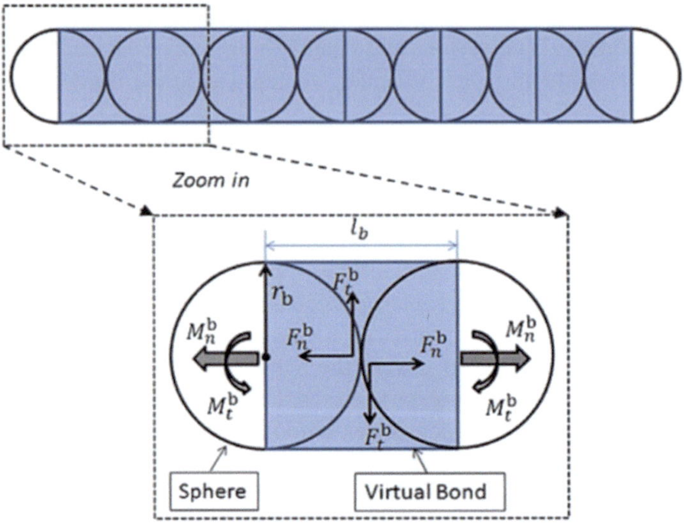

Fig. 4.2 Sketch map of the bonded-sphere model [2]

degrees of freedom associated with bond deformation require more sophisticated numerical integration schemes and smaller time steps to maintain solution stability. Contact detection becomes more complex as both sphere-sphere interactions and bond stress states must be monitored simultaneously. The model's computational overhead increases substantially with the number of active bonds, though modern parallel computing techniques and optimized contact algorithms have significantly improved its scalability for large-scale simulations.

Despite its advanced capabilities, the bonded-sphere model presents several practical challenges that require careful consideration. The parameters governing bond behavior demand comprehensive material characterization, often requiring iterative calibration procedures. The model's sensitivity to bond parameter selection can lead to significantly different macroscopic behaviors, necessitating thorough validation against experimental observations. Additionally, while the method excels at modeling particle-scale fracture and deformation, it may not capture certain bulk material properties as accurately as continuum-based approaches, particularly for highly dense or compacted systems.

4.1.3 Single Sphere with Rolling Resistance Model

The single sphere model with rolling resistance provides an efficient computational approach for approximating non-spherical particle behavior in DEM simulations while maintaining the fundamental simplicity of spherical particle representations. This model introduces an artificial resistance torque that opposes particle rotation, effectively emulating the geometric hindrance characteristic of non-spherical particles without explicitly modeling their complex shapes. The resistance mechanism operates through a dimensionless rolling friction coefficient that scales with the normal contact force, creating a torque proportional to both the contact force and particle radius. This approach preserves the computational advantages of spherical contact detection while capturing essential non-spherical effects on bulk material behavior. The physical basis of the rolling resistance model stems from its ability to replicate energy dissipation mechanisms observed in real granular systems. When non-spherical particles interact, their geometric features induce additional resistance to rotation through mechanisms such as surface deformation, micro-slip at contact points and geometric interlocking. The model abstracts these complex physical processes into a single parameterized torque that acts at each contact point.

The model demonstrates particular value in large-scale industrial applications where computational efficiency outweighs the need for precise geometric representation. In bulk material handling systems such as conveyor transfers or silo discharges, where particle shape primarily affects flow patterns rather than individual particle dynamics, the single sphere approximation with rolling resistance provides sufficient accuracy while enabling simulations of industrial-scale particle counts. The approach also proves useful in parametric studies and design optimization, where numerous simulation runs are required to explore operational conditions.

However, the model's simplifications introduce several inherent limitations that constrain its applicability. Most significantly, it cannot capture orientation-dependent phenomena or true geometric interactions between particles, as all contacts remain idealized sphere-sphere interactions. The resistance torque represents an averaged effect rather than physically precise behavior, which may lead to inaccuracies in applications where specific particle shape characteristics dominate the system dynamics. These limitations become particularly apparent in systems with highly anisotropic particles or where interlocking effects significantly influence bulk behavior. In such cases, more sophisticated shape representation methods are necessary to achieve accurate predictions.

4.2 Super-Ellipsoid-Based Model

The super-ellipsoid-based model originates from the superquadric equation, initially developed for computer graphics, and extends the quadric equation to represent a wide variety of particle shapes with smooth surfaces and sharp edges. Figure 4.3 displays the particles described by the super-ellipsoid model. The standard superquadric equation in 3D is defined as:

$$f(x, y, z) = \left(\left| \frac{x}{a} \right|^{S_2} + \left| \frac{x}{b} \right|^{S_2} \right)^{\frac{s_1}{s_2}} + \left| \frac{z}{c} \right|^{s_1} - 1 = 0, \tag{4.1}$$

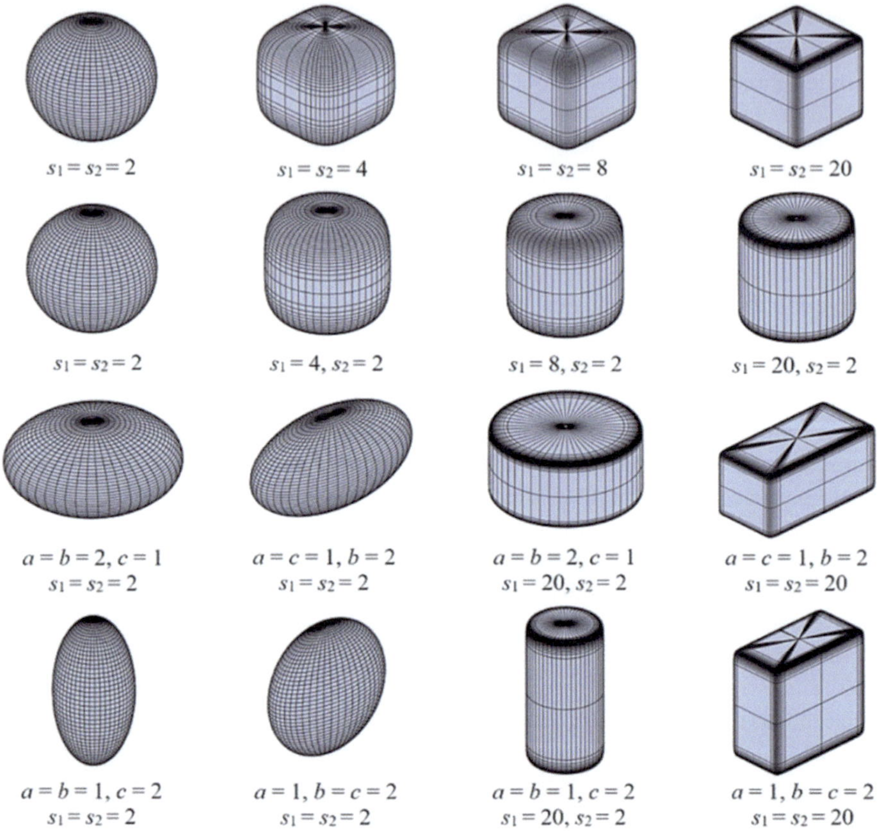

Fig. 4.3 Particles described by the super-ellipsoid model [1]

where a, b and c are the semi-axes lengths along the x, y and z directions, respectively. The shape indices, s_1 and s_2, decide the curvature of the particle edges, and the sharpness of the curvature increases with the shape indices.

The most notable feature of the super-ellipsoid-based model is its shape flexibility, allowing it to generate smooth, rounded forms like spheres and ellipsoids, as well as sharp-edged shapes such as cubes and cylinders, simply by adjusting the semi-axis lengths and shape indices. This adaptability makes it suitable for modeling many industrial particles, including pharmaceutical tablets, biomass pellets and granular materials. Unlike the multisphere model, which approximates shapes through clustered sub-spheres and often results in artificial surface roughness, the super-ellipsoid model provides a smooth and continuous surface representation, improving the realism of particle interactions in simulations. Another critical feature is its computational efficiency. While the model is generally faster than the multisphere approach for smooth particles, its performance depends on the particle shape. For instance, particles with extreme aspect ratios or high shape indices require more computational effort due to the complexity of contact detection. However, for most practical applications, the super-ellipsoid model strikes a balance between accuracy and computational cost.

The super-ellipsoid model can be implemented in different ways depending on the complexity of the particle shapes being simulated. The single-particle approach is the simplest form, where each particle is represented by a single super-ellipsoid. This method is ideal for convex and smoothly shaped particles, such as ellipsoids or rounded tablets, where a single mathematical expression suffices to define the geometry. For more complex or concave shapes, the composite particle approach is employed. One such variant is the multisuper-ellipsoid model, where a particle is constructed by combining multiple super-ellipsoids with overlapping regions. This allows the representation of irregular or asymmetrical particles, such as certain pharmaceutical tablets or fractured granules. Another variant, the poly-super-ellipsoid model, uses eight super-ellipsoid segments arranged in different quadrants to create asymmetric yet smooth particles. This method is particularly useful for particles that require precise control over local curvatures while maintaining overall smoothness.

The super-ellipsoid model has been widely adopted in CFD-DEM simulations to study particle–fluid interactions in various industrial and scientific contexts. In fluidization processes, it has been used to investigate the behavior of rod-like or disk-like particles in bubbling and spout-fluid beds, where particle orientation and drag forces play a critical role. The model's ability to capture shape-dependent phenomena, such as preferential alignment in flow fields, makes it valuable for optimizing fluidized bed reactors and mixers. In pneumatic and hydraulic conveying, the super-ellipsoid model helps analyze how non-spherical particles behave in pipelines, including their orientation, segregation tendencies and erosion effects on walls. This is particularly relevant in industries such as mining, agriculture and pharmaceuticals, where particle shape influences transport efficiency and equipment wear. Another key application is in pharmaceutical manufacturing, where the model is used to simulate processes like tablet coating, granulation and powder mixing. By accurately representing tablet shapes, the super-ellipsoid model enables researchers to study coating uniformity, particle collisions and segregation dynamics in Wurster coaters and other processing equipment.

4.3 Polyhedron-Based Model

Unlike approximation methods such as the multisphere or super-ellipsoid approaches, the polyhedron-based model precisely defines particle geometry through a collection of planar faces, edges and vertices, as shown in Fig. 4.4, making it particularly valuable for simulating angular particles commonly encountered in industrial and natural processes.

One of the model's most significant advantages lies in its ability to accurately capture the exact geometric features of particles, including sharp edges, flat faces and concavities. This level of precision makes it indispensable for applications where

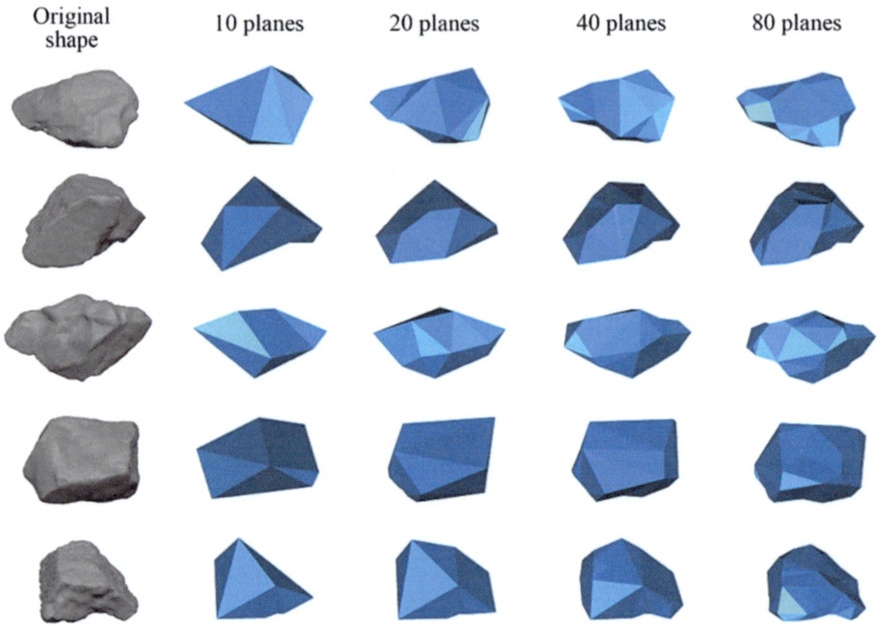

Fig. 4.4 Particles described by polyhedral model [1]

particle shape critically influences mechanical behavior, such as the interlocking of angular rocks in geotechnical applications. The model's accuracy stems from its direct representation of particle surfaces without the smoothing artifacts inherent in other approximation methods.

The polyhedron model exists in several variants to address different simulation requirements. Convex polyhedrons, including simple shapes like tetrahedrons and cubes, offer computational efficiency while maintaining geometric fidelity for many applications. For more complex particle morphologies, concave polyhedrons can be represented through decomposition into multiple convex sub-polyhedrons, though this approach increases computational demands. A particularly useful variant is the dilated polyhedron model, which modifies sharp edges by incorporating spherical caps or cylindrical bridges. This modification enhances numerical stability in dense granular systems while preserving the essential geometric characteristics of the particles.

Despite its geometric advantages, the polyhedron model presents certain computational challenges that must be considered. The complex contact detection algorithms required for polyhedral interactions result in significantly higher computational costs compared to simpler particle models. This makes the approach less efficient for systems dominated by smooth or rounded particles, where alternative models might be more appropriate. Additionally, the model requires careful calibration of contact parameters, such as stiffness and damping coefficients, to ensure physically realistic simulations.

4.4 Level Set Function Model

The level set function model is a surface representation method which describes particle boundaries through a continuous scalar field that assigns to every point in space a signed distance value relative to the nearest particle surface. Figure 4.5 displays the illustration of a level set function. The zero level set where the function equals zero precisely defines the particle boundary, while positive and negative values indicate positions outside and inside the particle, respectively.

The level set approach offers several distinctive benefits for DEM simulations. Its implicit representation eliminates the need for explicit surface meshing, avoiding the computational complexities associated with maintaining consistent surface tessellations during particle motion and deformation. The continuous scalar field enables sub-voxel precision in contact detection, significantly improving accuracy compared

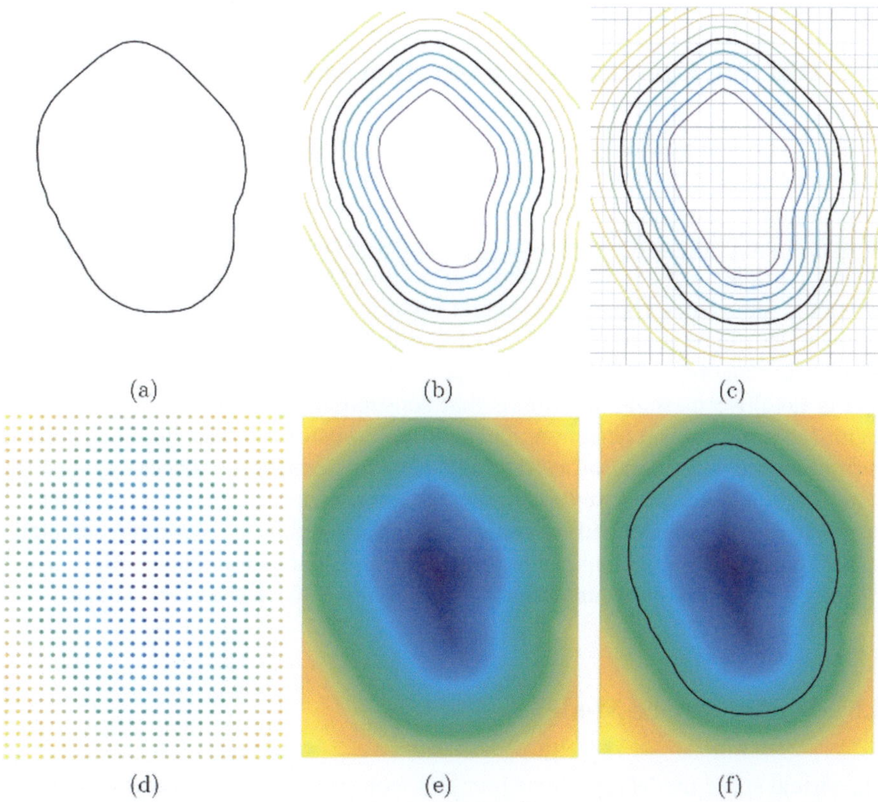

Fig. 4.5 Illustration of a level set function: **a** particle surface; **b** contour lines representing signed distance from surface; **c** superimposition on grid; **d** discretized level set function; **e** level set function with interpolation between grid points; **f** reconstruction of original grain surface via interpolation [3]

to discrete voxel-based methods. Furthermore, the model can represent both convex and concave features with equal fidelity, including undercuts and internal voids that challenge conventional surface-based representations. While powerful, the level set method introduces notable computational demands. Memory requirements scale with the resolution of the level set grid, and the need to maintain the signed distance property requires periodic reinitialization of the scalar field.

Implementing the level set model in DEM requires specialized algorithms for several critical operations. Contact detection between particles involves finding the minimum distance between their respective zero level sets, typically solved through narrow-band optimization techniques. Force calculations leverage the gradient information from the level set field to determine contact normals and penetration depths. For computational efficiency, the level set function is usually discretized on a uniform grid surrounding each particle, with careful attention paid to maintaining the signed distance property during particle motion through reinitialization procedures.

4.5 Discrete Function Representation Model

Unlike traditional methods that rely on analytical shape descriptions, discrete function representation (DFR) model represents particle surfaces through discrete computational points, as shown in Fig. 4.6, offering flexibility in handling irregular, non-convex or experimentally measured particle morphologies. DFR works by decomposing particle surfaces into a network of discrete nodes, each storing critical geometric information including spatial coordinates and surface normal vectors. These nodes can be generated through various techniques: mathematical sampling of parametric equations for defined shapes, point cloud reconstruction from 3D scanning data or adaptive meshing algorithms that concentrate resolution in high-curvature regions. During contact detection, the algorithm transforms nodes from one particle into the coordinate frame of another and performs penetration checks against the target particle's boundary representation. When a node from particle i is found to penetrate the volume defined by particle j's surface nodes, the algorithm calculates contact parameters including penetration depth and contact normal direction, which are then used to compute interaction forces based on selected contact models.

4.6 Virtual Space Model

The virtual space model represents particles with complex geometries by digitizing them into a collection of pixels in 2D or voxels in 3D, as shown in Fig. 4.7. This method discretizes the particle's volume or surface into a grid-based structure, where each pixel or voxel acts as a binary marker indicating whether a specific region of space is occupied by the particle. By converting continuous shapes into discrete spatial units, the virtual space model enables the simulation of arbitrarily intricate

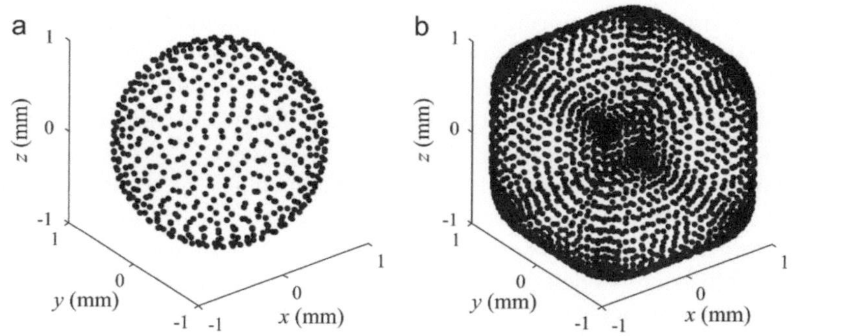

Fig. 4.6 Discrete function representation of particles: **a** spherical particle (uniform discretization) and **b** super-quadric particle (adaptive discretization) [4]

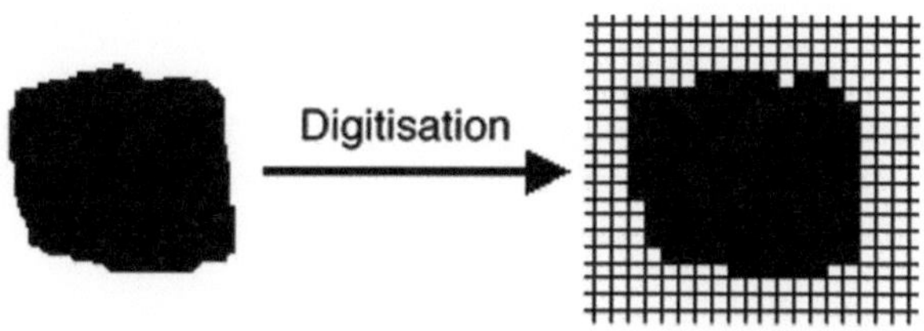

Fig. 4.7 Digitization of a real object [5]

particle morphologies, including those with concave features, internal pores or irregular surfaces that are challenging to describe using traditional analytical methods like spheres or ellipsoids. In this model, particles are constructed by assigning a value of "1" to voxels or pixels that lie within the particle's boundary and "0" to those outside. The resolution of the grid directly influences the accuracy of the geometric representation, that is finer grids capture more detail but require greater computational resources. The shape data can be derived from mathematical definitions such as discretized superquadrics, or experimental measurements such X-ray tomography and 3D scanning of real particles.

A critical advantage of the voxel-based approach is its ability to accurately reproduce intricate surface features and internal structures that would be challenging or impossible to represent using conventional analytical shape models. The method naturally accommodates surface roughness, internal porosity and truly arbitrary particle contours without requiring simplifying geometric assumptions. This fidelity comes at the cost of increased computational complexity, as each particle's behavior must be computed through the collective dynamics of its constituent voxels. Contact detection between particles involves identifying overlapping voxel regions and calculating interaction forces based on the penetration depth and material properties of the contacting voxels.

Despite its powerful capabilities, the virtual space model faces several limitations that constrain its widespread adoption. The computational cost remains substantially higher than for analytical shape models, particularly for systems containing large numbers of particles or requiring high voxel resolutions. Memory requirements grow quickly with increasing system size and resolution, necessitating careful optimization of data storage formats. The discrete nature of voxel representations can introduce artificial roughness in contact force calculations, potentially affecting simulation accuracy.

Overall, the choice of particle shape representation in DEM depends on the trade-off between computational efficiency and the accuracy of the shape approximation. Spherical particles offer maximum computational efficiency but poor shape realism. Multisphere models provide better shape approximation while maintaining reasonable computational cost, though they create artificially rough surfaces. Super-ellipsoids efficiently represent smooth, regular shapes like rods or tablets but cannot handle complex geometries. For higher accuracy, polyhedral models precisely capture angular shapes but require intensive contact calculations. Virtual space models handle arbitrary geometries including porous structures, but at high computational cost. Level set methods enable dynamic shape changes but are computationally demanding. Discrete function representation uses surface nodes for flexible shape modeling with moderate efficiency. Other specialized approaches include non-uniform rational B-Splines for smooth surfaces, Minkowski sums for composite shapes and Boolean operations for constructed geometries.

4.7 Contact Detection Algorithms

4.7.1 Geometric Potential Algorithm

The geometric potential (GP) algorithm treats particle surfaces as equipotential contours in a scalar field, as shown in Fig. 4.8. This mathematical framework conceptualizes each particle as generating a potential field in space, where the particle surface corresponds to a specific isopotential contour. When two particles approach each other, their potential fields interact, and contact is identified by analyzing the combined field to locate points of minimum potential energy between the surfaces. This method demonstrates particular effectiveness when applied to smoothly varying particle geometries, especially those represented by superquadric models or level set functions. The continuous nature of these shape representations aligns perfectly with the potential field approach, allowing for mathematically consistent contact resolution.

4.7.2 Common Normal Algorithm

The common normal (CN) algorithm operates on the principle that the contact force between two objects acts along the shortest connecting line that is perpendicular

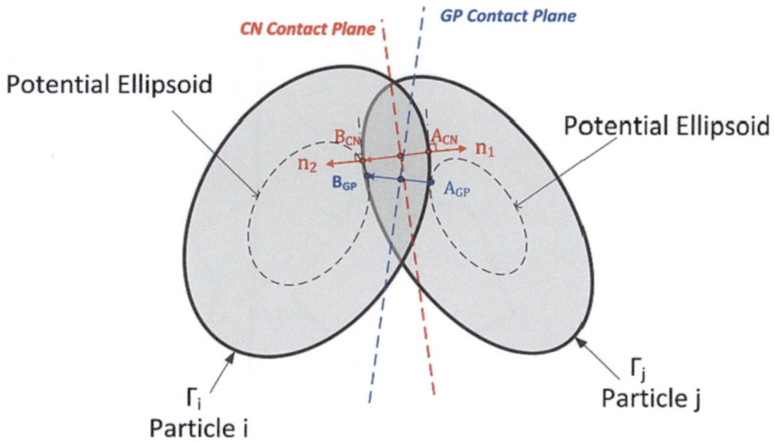

Fig. 4.8 Schematic illustration of the contact parameters determined by GP and CN method [6]

to both surfaces at the contact location, as shown in Fig. 4.8. This line, known as the common normal, defines the direction of the repulsive force and is critical for accurately modeling physical interactions in DEM simulations. For simple shapes like spheres, the common normal is straightforward to compute, as it coincides with the line connecting their centers. However, for more complex geometries such as polyhedral and superquadrics particles, the algorithm must account for local surface curvatures and closest-point projections.

4.7.3 Common Plane Algorithm

The common plane (CP) algorithm seeks to determine a virtual plane positioned between two particles that satisfies specific geometric criteria, as shown in Fig. 4.9. For convex particles that are not penetrating, such a plane completely separates the two shapes. When particles overlap, the algorithm finds the plane that minimizes some measure of their mutual penetration. The contact normal vector is then derived from this common plane's orientation, while the penetration depth is calculated based on the plane's position relative to the particles' surfaces. The algorithm proceeds through an iterative optimization process that typically begins with an initial plane estimate based on the particles' centroids and orientations. Through successive refinements, the method adjusts the plane's position and orientation to minimize either the separation distance for non-contacting particles or the penetration measure for overlapping particles. This optimization ensures robust contact detection even for particles with complex faceted geometries.

Key advantages of the CP algorithm include its mathematical rigor in handling polyhedron-polyhedron interactions and its ability to provide consistent contact

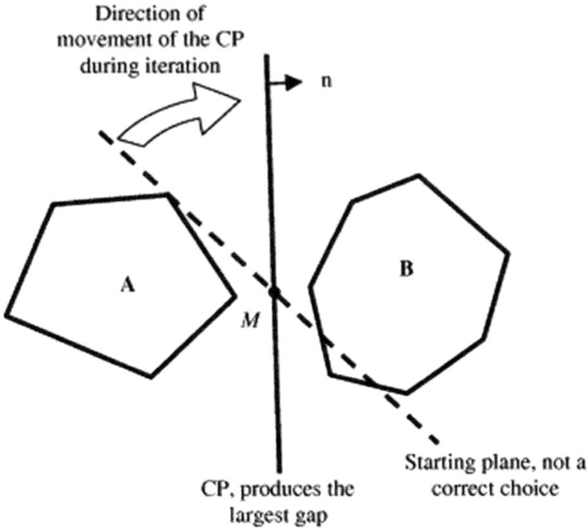

Fig. 4.9 Conventional iterative procedure for finding the CP [7]

normals. The method naturally accommodates various contact configurations, including vertex-face, edge-edge and face-face contacts that commonly occur between angular particles. Its formulation guarantees that for non-overlapping convex polyhedrons, at least one such separating plane exists, making the algorithm particularly reliable for many engineering applications involving crushed rock or other fragmented materials. In practical implementation, the CP algorithm demonstrates particular strength when applied to convex polyhedral particles, where it can efficiently resolve contacts with well-defined geometric properties. Its computational efficiency compares favorably to other polyhedron contact detection methods, especially when combined with spatial partitioning techniques to reduce the number of potential contact pairs. However, the algorithm faces challenges when applied to highly concave particles, as the fundamental separating plane concept becomes less straightforward. In such cases, the concave particles can be decomposed into convex components and the CP algorithm can be applied to each sub-part.

In addition to the common GP, CN and CP algorithms, other contact detection methods for non-spherical particles in DEM include the Gilbert-Johnson-Keerthi algorithm, expanding polytope algorithm, Minkowski difference-based methods, surface-based approaches like the signed distance function and grid-based or hierarchical bounding volume techniques. Machine learning-based methods are also emerging for complex shapes. These algorithms vary in accuracy, computational efficiency and applicability to different particle geometries. The choice depends on the trade-off between precision and computational cost, with Gilbert-Johnson-Keerthi and expanding polytope algorithms being popular for convex shapes, while bounding volume hierarchies or grid-based methods are often used for scalability in large systems. For non-convex particles, decomposition into convex subparts may be preferred.

References

1. H. Ma, L. Zhou, Z. Liu, M. Chen, X. Xia, Y. Zhao, A review of recent development for the CFD-DEM investigations of non-spherical particles. Powder Technol. **412**, 117972 (2022). https://doi.org/10.1016/j.powtec.2022.117972
2. Y. Guo, C. Wassgren, B. Hancock, W. Ketterhagen, J. Curtis, Validation and time step determination of discrete element modeling of flexible fibers. Powder Technol. **249**, 386–395 (2013). https://doi.org/10.1016/j.powtec.2013.09.007
3. R. Kawamoto, E. Andò, G. Viggiani, J.E. Andrade, Level set discrete element method for three-dimensional computations with triaxial case study. J. Mech. Phys. Solids **91**, 1–13 (2016). https://doi.org/10.1016/j.jmps.2016.02.021
4. G. Lu, J.R. Third, C.R. Müller, Critical assessment of two approaches for evaluating contacts between super-quadric shaped particles in DEM simulations. Chem. Eng. Sci. **78**, 226–235 (2012). https://doi.org/10.1016/j.ces.2012.05.041
5. X. Jia, R.A. Williams, A packing algorithm for particles of arbitrary shapes. Powder Technol. **120**(3), 175–186 (2001). https://doi.org/10.1016/S0032-5910(01)00268-6
6. K. Kildashti, K. Dong, B. Samali, Q. Zheng, A. Yu, Evaluation of contact force models for discrete modelling of ellipsoidal particles. Chem. Eng. Sci. **177**, 1–17 (2018). https://doi.org/10.1016/j.ces.2017.11.004
7. E.G. Nezami, Y.M.A. Hashash, D. Zhao, J. Ghaboussi, A fast contact detection algorithm for 3-D discrete element method. Comput. Geotech. **31**(7), 575–587 (2004). https://doi.org/10.1016/j.compgeo.2004.08.002

Chapter 5
Computational Fluid Dynamics

5.1 Governing Equations

The foundation of computational fluid dynamics (CFD) lies in the governing equations that describe fluid motion. The equations derived from the principles of conservation of mass and momentum are expressed in mathematical forms that account for the dynamic behaviors of fluids.

5.1.1 Continuity Equation

The continuity equation represents the mathematical expression of mass conservation in a flowing continuum, which is expressed as:

$$\frac{\partial\left(\alpha_f \rho_f\right)}{\partial t} + \nabla \cdot \left(\alpha_f \rho_f \boldsymbol{v}_f\right) = 0 \tag{5.1}$$

where ρ denotes the fluid density, $\boldsymbol{v}_f$ represents the fluid velocity, α_f is the porosity or void fraction. The first term captures the temporal change in fluid mass in a control volume, while the second term accounts for the net convective flux of mass through the control volume boundaries.

5.1.2 Momentum Equation

The Navier–Stokes equations constitute the momentum transport equations derived from Newton's second law applied to a fluid continuum, which is expressed as:

R. K. Agarwal and Y. Shao, *Theory and Application of Discrete Element Method to Simulation of Multiphase Flows*, https://doi.org/10.1007/978-3-032-11157-9_5

$$\frac{\partial\left(\alpha_f \rho_f \boldsymbol{v}_f\right)}{\partial t} + \nabla \cdot \left(\alpha_f \rho_f \boldsymbol{v}_f \otimes \boldsymbol{v}_f\right) = -\alpha_f \nabla p + \nabla \cdot \alpha_f \boldsymbol{\tau}_f + \alpha_f \rho_f \boldsymbol{g} + \boldsymbol{F}_{pf} \tag{5.2}$$

where $\boldsymbol{\tau}_f$ represents the viscous stress tensor, p is the fluid pressure, $\boldsymbol{g}$ is the gravitational acceleration, and $\boldsymbol{F}_{pf}$ is the particle–fluid interaction term. The left-hand side terms represent the substantial derivative of momentum change, comprising the rate of local momentum change and the momentum flux due to fluid motion. The right-hand side includes the pressure gradient force, viscous forces, gravitational body force and additional body forces that become particularly relevant in CFD-DEM coupling for representing fluid-particle interactions.

$$\boldsymbol{F}_{pf} = \boldsymbol{F}_d + \boldsymbol{F}_{lift} + \boldsymbol{F}_{vm} + \boldsymbol{F}_{basset} + \boldsymbol{F}_{other} \tag{5.3}$$

where $\boldsymbol{F}_d$ is the drag force; $\boldsymbol{F}_{lift}$ is the lift force including the Saffman lift due to shear gradients and Magnus lift from particle rotation; $\boldsymbol{F}_{vm}$ is the virtual mass accounting for the fluid inertia around accelerating particles; $\boldsymbol{F}_{basset}$ is the Basset history force which represents viscous diffusion effects. The additional force F_{other} may include electrostatic forces in charged systems, magnetic forces in magnetorheological fluids and thermophoretic effects in non-isothermal flows.

5.2 Particle–Fluid Interaction Forces

In CFD-DEM simulations, the interaction forces between particles and fluid play a critical role in accurately capturing the dynamics of particle–fluid systems. Among these forces, the drag force is the most significant, as it arises due to the relative motion between the particles and the surrounding fluid. Other interaction forces, such as the lift force and virtual mass force may also be considered depending on the specific application, though their contributions are often secondary compared to the drag force.

5.2.1 Drag Force

The drag force is the dominant particle–fluid interaction force and is caused by the resistance experienced by a particle moving through a fluid. For an isolated particle, the drag force F_{d0} can be calculated using the following equation:

$$\boldsymbol{F}_{d0} = \frac{1}{2} C_D \rho_f \frac{\pi d_p^2}{4} \left(\boldsymbol{v}_f - \boldsymbol{v}_p\right) \left|\boldsymbol{v}_f - \boldsymbol{v}_p\right| \tag{5.4}$$

where d_p is the diameter of the volume-equivalent sphere; $\boldsymbol{v}_f$ is the fluid velocity; $\boldsymbol{v}_p$ is the particle velocity; C_D is the drag coefficient, which is influenced by factors such as particle shape, orientation and local flow conditions.

The drag coefficient C_D derived from Schiller-Naumann (1933) model is expressed as [1]:

$$C_D = \begin{cases} \frac{24}{Re}\left(1 + 0.15Re_p^{0.687}\right), Re_p < 1000 \\ 0.44, Re_p \geq 1000 \end{cases} \tag{5.5}$$

$$Re_p = \frac{\rho_f \left|\boldsymbol{v}_f - \boldsymbol{v}_p\right| d_p}{\mu_f} \tag{5.6}$$

where μ_f is the fluid dynamic viscosity.

However, for dense granular systems such as fluidized beds, the presence of neighboring particles alters the flow field around each particle, requiring corrections to the drag force. This is often accounted for by introducing the void fraction α_f, which represents the proportion of fluid in a given volume. The general form of corrected drag force F_d for dense granular systems is expressed as:

$$F_d = F_{d0} f\left(\alpha_f\right) \tag{5.7}$$

where $f\left(\alpha_f\right)$ is a function of the void fraction, often derived empirically or through theoretical models.

(1) Wen and Yu drag model

The Wen and Yu (1966) model is a widely used empirical correlation for calculating the gas–solid drag force in dilute to moderately dense particle suspensions [2], which is typically used for the conditions with particle volume fractions α_p smaller than 0.2 ($\alpha_p + \alpha_f = 1$). The drag force per unit volume in a particle–fluid mixture is:

$$\boldsymbol{F}_d = \beta\left(\boldsymbol{v}_f - \boldsymbol{v}_p\right) \tag{5.8}$$

where β is the interphase momentum exchange coefficient, expressed as:

$$\beta = \frac{3}{4} C_D \frac{\rho_f \left(1 - \alpha_p\right)\left|\boldsymbol{v}_f - \boldsymbol{v}_p\right|}{d_p} \alpha_p^{-2.65} \tag{5.9}$$

(2) Gidaspow drag model

For higher solid fractions, the Wen & Yu model becomes inaccurate. Instead, the Gidaspow (1994) hybrid model is used [3], which combines the Wen & Yu model for dilute flows and the Ergun equation for dense flows.

For dilute to moderate flows with $\alpha_p < 0.2$, the interphase momentum exchange coefficient β in the Gidaspow model is obtained from the Wem & Yu model:

$$\beta = \frac{3}{4} C_D \frac{\rho_f (1 - \alpha_p) |\mathbf{v}_f - \mathbf{v}_p|}{d_p} \alpha_p^{-2.65} \tag{5.10}$$

Conversely, for $\alpha_p \geq 0.2$, the exchange coefficient β proposed by Ergun is used as:

$$\beta = 150 \frac{(1 - \alpha_p)^2}{\alpha_p} \frac{\mu_f}{d_p^2} + 1.75 \frac{(1 - \alpha_p)}{\alpha_p} \frac{\rho_f}{d_p} |\mathbf{v}_f - \mathbf{v}_p| \tag{5.11}$$

(3) Syamlal and O'Brien drag model

The Syamlal-O'Brien drag model (1987) is derived from experimental sedimentation data and provides a particle-terminal-velocity-dependent formulation [4], which is expressed as.

$$\beta = \frac{3}{4} C_D \frac{\rho_f (1 - \alpha_p) |\mathbf{v}_f - \mathbf{v}_p|}{d_p} \frac{(1 - \alpha_p)}{\alpha_f^2 v_{r,s}^2} \tag{5.12}$$

where $v_{r,s}$ is the dimensionless slip velocity.

$$v_{r,s} = 0.5\left(A - 0.06 Re_p + \sqrt{(0.06 Re_p)^2 + 0.12 Re_p (2B - A) + A^2} \right) \tag{5.13}$$

$$A = \alpha_f^{4.14} \tag{5.14}$$

$$B = \begin{cases} 0.8\alpha_f^{1.28}, \alpha_f \leq 0.85 \\ \alpha_f^{2.65}, \alpha_f > 0.85 \end{cases} \tag{5.15}$$

(4) Subgrid drag model

Subgrid drag models are advanced corrections applied to traditional drag laws in gas–solid flow simulations, designed to account for unresolved microscale phenomena such as particle clustering, local voidage fluctuations and turbulence effects. These models are critical in coarse-grid simulations, where the grid size is larger than the characteristic scales of particle aggregates or bubbles. Without subgrid corrections, conventional drag models such as Wen & Yu or Gidaspow tend to overpredict drag in clustered regions and underpredict it in dilute zones, leading to inaccurate predictions of flow dynamics.

Several prominent subgrid drag models have been developed to tackle these challenges. The model proposed by Igci et al. [5] uses the direct numerical simulation-based filtering to derive correction factors as functions of local solid volume fraction and Reynolds number. The Energy Minimization Multiscale (EMMS) model, originally developed by Li and Kwauk [6], employs an energy optimization principle to determine drag reduction due to clustering. Agrawal et al. [7] introduced a filtered approach that accounts for unresolved solid fraction fluctuations, while Parmentier et al. [8] developed a structure-based model correlating drag with cluster packing density. Each model offers unique advantages: EMMS excels in capturing cluster-induced drag reduction, whereas filtered models like Agrawal's are simpler to implement but require calibration.

For non-spherical particles, the drag force calculation becomes more complex due to the dependence on particle shape and orientation. General correlations for the drag coefficient C_D often incorporate shape factors, such as sphericity or aspect ratio, to account for these effects.

5.2.2 *Lift Force*

The lift force arises due to velocity gradients in the fluid or particle rotation. Unlike drag forces that resist motion along the flow direction, lift forces cause particles to migrate across streamlines, leading to phenomena like particle segregation, preferential orientation and non-uniform spatial distribution. For spherical particles, two primary mechanisms generate lift forces. The Saffman lift force occurs due to velocity gradients in the shear flow, where particles experience different fluid velocities on opposite sides. This force is especially relevant near walls or in regions with strong shear. The Magnus lift force results from particle rotation, which creates asymmetric pressure distributions around the particle surface. Both forces are well-characterized for spherical particles at low-to-moderate Reynolds numbers, with established mathematical formulations that depend on particle size, fluid viscosity, and local flow conditions. The general expression for the lift force F_{lift} acting on a particle in a fluid flow is commonly written as:

$$F_{lift} = 0.5 C_L \rho_f \frac{\pi}{4} d_p^2 \left| \boldsymbol{v}_f - \boldsymbol{v}_p \right|^2 \tag{5.16}$$

where C_L is the lift coefficient.

To calculate the lift force accurately, the evaluation of the lift coefficient C_L is very important which is influenced by particle shape, orientation and flow conditions. Compared with spherical particles, non-spherical particles exhibit more complex lift behavior due to their shape-dependent interactions with the fluid flow. For elongated particles like fibers or rods, the lift force can cause preferential alignment where particles orient themselves to minimize drag resistance. Disk-shaped particles, on the other hand, may experience lateral migration across streamlines due to asymmetric pressure distributions. Researchers have developed specialized lift models for common non-spherical shapes such as ellipsoids and spherocylinders, incorporating parameters like crosswise and lengthwise sphericity to account for orientation effects.

5.2.3 *Other Forces*

Besides drag and lift forces, several other important interaction forces influence particle dynamics in fluid-particle systems. These forces become particularly significant in specific flow regimes or for certain particle characteristics. The virtual mass force accounts for the additional inertia required to accelerate the surrounding fluid when a particle changes velocity. This force becomes significant when the particle density is comparable to the fluid density, such as in liquid–solid systems. This force is calculated using a virtual mass coefficient that depends on particle shape and flow conditions. The Basset history force represents the temporal delay in boundary layer development during particle acceleration, particularly relevant for small particles in viscous fluids undergoing unsteady motion. Its computation involves a memory integral that captures the entire acceleration history of the particle. Thermophoretic forces, driven by temperature gradients in the fluid, cause particles to migrate from warmer to cooler regions, playing important roles in combustion systems and thermal precipitators. For charged particles in electric fields, electrostatic forces including Coulomb forces and dielectrophoresis can dominate particle behavior, with applications ranging from powder coating to electrostatic precipitators.

5.3 Turbulence Modeling

In the momentum equation, the term $\boldsymbol{\tau}_f$ represents the viscous stress tensor of the fluid phase, which accounts for the internal frictional forces arising due to fluid viscosity and velocity gradients. In continuum mechanics, the viscous stress tensor is a symmetric second-order tensor that describes the deviatoric part of the total stress, capturing the resistance to deformation under shear or extensional flows. For a Newtonian fluid, $\boldsymbol{\tau}_f$ is typically modeled using Stokes' hypothesis, which relates it linearly to the rate of strain tensor through the fluid's dynamic viscosity. The equation is expressed as:

$$\boldsymbol{\tau}_f = \alpha_f \mu_f \left(\nabla \boldsymbol{v}_f + \nabla \boldsymbol{v}_f^T - \frac{2}{3} (\nabla \cdot \boldsymbol{v}_f) \boldsymbol{I} \right) \tag{5.17}$$

where μ_f is the dynamic viscosity of the fluid.

While the viscous stress tensor provides a complete description for laminar flows of Newtonian fluids, most practical engineering applications involve turbulent flow regimes. To address this challenge, three primary modeling approaches have been developed, including Reynolds-Averaged Navier–Stokes (RANS) simulation which models all turbulent effects through statistical averaging, Large Eddy Simulation (LES) which resolves large scales while modeling small scales, and direct numerical simulation (DNS) which resolves all turbulent scales.

5.3.1 Reynolds-Averaged Navier–Stokes (RANS) Simulation

The RANS methodology decomposes flow variables into mean and fluctuating components, allowing for the statistical modeling of turbulence effects. In the RANS framework, the instantaneous velocity $\boldsymbol{v}_{fi}$ is split into a time-averaged component $\boldsymbol{v}_f$ and a fluctuating component $\boldsymbol{v}_f\prime$, such that $\boldsymbol{v}_{fi} = \boldsymbol{v}_f + \boldsymbol{v}_f\prime$. Similarly, the phase fraction α_f may also be decomposed if it exhibits significant fluctuations, though it is often treated as a steady field in many applications. Applying Reynolds averaging to the viscous stress tensor $\boldsymbol{\tau}_f$ yields the mean viscous stress $\boldsymbol{\tau}_f$, which retains the same functional form but operates on the mean velocity gradient. However, the averaging process introduces additional unclosed terms, most notably the Reynolds stress tensor τ_f^{Re}, which represents the turbulent momentum transport due to velocity fluctuations. To close the system of equations, the Reynolds stress must be modeled. The most common approach is the Boussinesq eddy-viscosity hypothesis, which assumes that the turbulent stresses are proportional to the mean strain rate tensor, which is expressed as:

$$\tau_f^{\mathrm{Re}} = 2\alpha_f \mu_t \left(\boldsymbol{S} - \frac{1}{3}(\nabla \cdot \boldsymbol{v}_f)\boldsymbol{I} \right) - \frac{2}{3}\alpha_f \rho_f k \boldsymbol{I} \tag{5.18}$$

where μ_t is the fluid turbulent viscosity; $\boldsymbol{S}$ is the mean strain rate tensor; k is the turbulent kinetic energy. The turbulent viscosity is typically modeled using different models, such as the standard k-ε model, Shear Stress Transport (SST) k-ω model, Wray-Agarwal (WA) model and Spalart–Allmaras (SA) model.

(1) Standard k-ε turbulence model

As a semi-empirical model, the standard k-ε turbulence model [9] is described by transport equations for the turbulence kinetic energy k and its dissipation rate ε:

$$\frac{\partial(\alpha_f \rho_f k)}{\partial t} + \nabla \cdot (\alpha_f \rho_f \boldsymbol{v}_f k) = \nabla \cdot \left(\alpha_f \frac{\mu_t}{\sigma_k} \nabla k \right) + \alpha_f (G_k - \rho_f \varepsilon) + \alpha_f \rho_f \Pi_k \tag{5.19}$$

$$\frac{\partial(\alpha_f \rho_f \varepsilon)}{\partial t} + \nabla \cdot (\alpha_f \rho_f \boldsymbol{v}_f \varepsilon) = \nabla \cdot \left(\alpha_f \frac{\mu_t}{\sigma_\varepsilon} \nabla \varepsilon \right) + \alpha_f \left(C_{1\varepsilon} \frac{\varepsilon}{k} G_k - C_{2\varepsilon} \rho_f \frac{\varepsilon^2}{k} \right) + \alpha_f \rho_f \Pi_\varepsilon \tag{5.20}$$

where G_k denotes the generation of k. Π_k and Π_ε represent the interphase turbulent exchange. $C_{1\varepsilon}$ and $C_{2\varepsilon}$ are constants with the values of 1.44 and 1.92 respectively. σ_k and σ_ε are turbulent Prandtl numbers for k and ε respectively, with $\sigma_k = 1.0$, $\sigma_\varepsilon = 1.3$.

The eddy-viscosity is computed as a function of k and ε:

$$\mu_t = \rho_f C_\mu \frac{k^2}{\varepsilon} \tag{5.21}$$

where C_μ is a constant with value 0.09.

(2) SST k-ω turbulence model

As a two-equation model, the SST k-ω turbulence model [9] was developed by Menter on the basis of the transport equations for turbulent kinetic energy k and the specific dissipation rate ω. The transport equations are given by:

$$\frac{\partial(\alpha_f \rho_f k)}{\partial t} + \nabla \cdot (\alpha_f \rho_f \mathbf{v}_f k) = \nabla \cdot \left[\alpha_f \left(\mu_l + \frac{\mu_t}{\sigma_k}\right)\nabla k\right] + \alpha_f G_k - \alpha_f \rho_f \beta^* \omega k \tag{5.22}$$

$$\frac{\partial(\alpha_f \rho_f \omega)}{\partial t} + \nabla \cdot (\alpha_f \rho_f \mathbf{v}_f \omega) = \nabla \cdot \left[\alpha_f \left(\mu_l + \frac{\mu_t}{\sigma_\omega}\right)\nabla \omega\right] + \alpha_f G_\omega - \alpha_f \rho_f \beta \omega^2 + 2(1 - F_1)\frac{\alpha_f \rho_f}{\sigma_{\omega,2}\omega}\nabla k \cdot \nabla \omega \tag{5.23}$$

where G_k and G_ω denote the production of k and ω respectively. The turbulent Prandtl numbers for k and ω are σ_k and σ_ω, respectively, which are expressed as follows:

$$\sigma_k = \frac{1}{F_1/\sigma_{k,1} + (1 - F_1)/\sigma_{k,2}} \tag{5.24}$$

$$\sigma_\omega = \frac{1}{F_1/\sigma_{\omega,1} + (1 - F_1)/\sigma_{\omega,2}} \tag{5.25}$$

$\sigma_{k,1}$,$\sigma_{k,2}$, $\sigma_{\omega,1}$ and $\sigma_{\omega,2}$ are constants with the values of 1.176, 1.0, 2.0 and 1.168, respectively.

The blending function F_1 is expressed as:

$$F_1 = \tan h\left\{\left\{\min\left[\max\left(\frac{\sqrt{\kappa}}{0.09\omega d}, \frac{500\mu}{\rho d^2 \omega}\right), \frac{4\rho_f k}{\sigma_{\omega,2} D_\omega^+ d^2}\right]\right\}^4\right\} \tag{5.26}$$

$$D_\omega^+ = \max\left(2\rho_g \frac{1}{\sigma_{\omega,2}}\frac{1}{\omega}\nabla k \cdot \nabla \omega,\ 10^{-10}\right) \tag{5.27}$$

where D_ω^+ represents the positive portion of the cross-diffusion term.

The eddy-viscosity μ_t for SST k-ω model is given by:

$$\mu_t = \frac{\rho_f k}{\omega}\frac{1}{\max\left[\frac{1}{\alpha^*}, \frac{SF_2}{a_1 \omega}\right]} \tag{5.28}$$

$$F_2 = \tanh\left\{\left[\max\left(2\frac{\sqrt{k}}{0.09\omega d}, \frac{500\mu}{\rho_f d^2 \omega}\right)\right]^2\right\} \tag{5.29}$$

where S is the magnitude of strain rate and a_1 is a constant = 0.31.

(3) Wray-Agarwal (WA) turbulence model

The WA turbulence model [10] is a one-equation model which is derived from the k-ω model to obtain a transport equation for R ($R = k/\omega$). WA 2017m version of the model is given below:

$$\begin{aligned}\frac{\partial R}{\partial t} + \nabla \cdot (\boldsymbol{v}_f R) &= \nabla \cdot \left[(\sigma_R R + \nu_f)\nabla R\right] + C_1 RS \\ &+ f_1 C_{2k\omega}\frac{R}{S}\nabla R \cdot \nabla S \\ &- (1 - f_1)C_{2k\varepsilon}\min[R^2\left(\frac{\nabla S \cdot \nabla S}{S^2}\right), C_m \nabla R.\nabla R]\end{aligned} \tag{5.30}$$

where ν_f is the kinematic viscosity; C_1, $C_{2k\omega}$, $C_{2k\varepsilon}$ and σ_R are constants; S represents the mean strain given below.

$$S = \sqrt{2S_{ij}S_{ij}}, \quad S_{ij} = \frac{1}{2}\left(\frac{\partial \boldsymbol{v}_{fi}}{\partial x_j} + \frac{\partial \boldsymbol{v}_{fj}}{\partial x_i}\right) \tag{5.31}$$

The switching function f_1 is expressed as:

$$f_1 = \min(\tanh(\arg_1^4), 0.9) \tag{5.32}$$

$$\arg_1 = \frac{1 + \frac{d\sqrt{RS}}{\nu}}{1 + \left[\frac{\max\left(d\sqrt{RS}, 1.5R\right)}{20\nu}\right]^2} \tag{5.33}$$

where d is the distance from the nearest wall.

The constants in the transport equation are given as:

$$\begin{aligned}C_{1k\omega} &= 0.0829 C_{1k\varepsilon} = 0.1127 \\ C_1 &= f_1(C_{1k\omega} - C_{1k\varepsilon}) + C_{1k\varepsilon} \\ \sigma_{k\omega} &= 0.72 \quad \sigma_{k\omega} = 1.0 \\ \sigma_R &= f_1(\sigma_{k\omega} - \sigma_{k\varepsilon}) + \sigma_{k\varepsilon} \\ k &= 0.41\end{aligned}$$

$$C_{2k\omega} = \frac{C_{1k\omega}}{k^2} + \sigma_{k\omega}, \; C_{2k\varepsilon} = \frac{C_{1k\varepsilon}}{k^2} + \sigma_{k\varepsilon} \tag{5.34}$$

The turbulence eddy-viscosity is expressed as:

$$\mu_t = \rho_f f_\mu R \tag{5.35}$$

$$f_{\mu} = \frac{\chi^3}{\chi^3 + C_{\omega}^3}, \chi = \frac{R}{\nu_f} \tag{5.36}$$

where f_{μ} is the damping function which describes the wall blocking effect and C_m is a constant with the value of 8.54.

(4) Spalart–Allmaras (SA) turbulence model

The Spalart–Allmaras (SA) turbulence model is an industry standard one-equation linear eddy-viscosity turbulence model developed for the prediction of aerodynamic flows [11]. It is a low-Reynolds number model, and requires the boundary layer near the wall for accuracy. The SA turbulence model is given by the following equation:

$$\frac{\partial \tilde{\nu}}{\partial t} + \boldsymbol{v}_f \cdot \nabla \tilde{\nu} = C_{b1} \tilde{S} \tilde{\nu} - C_{w1} f_w \left(\frac{\tilde{\nu}}{d} \right)^2 + \frac{1}{\sigma} \left[\nabla \cdot ((\nu + \tilde{\nu}) \nabla \tilde{\nu}) + C_{b2} (\nabla \tilde{\nu})^2 \right] \tag{5.37}$$

where $\tilde{\nu}$ is the modified turbulent kinematic viscosity.

The turbulent eddy-viscosity is given by the equation:

$$\mu_t = \rho_f \tilde{\nu} f_{v1} \tag{5.38}$$

Near wall blocking is accounted for by the damping function f_{v1}:

$$f_{v1} = \frac{\chi^3}{\chi^3 + C_{v1}^3} \tag{5.39}$$

$$\chi = \frac{\tilde{\nu}}{\nu} \tag{5.40}$$

The remaining functions are defined as follows:

$$\tilde{S} = \Omega + \frac{\tilde{\nu}}{k^2 d^2} f_{v2}, f_{v2} = 1 - \frac{\chi}{1 + \chi f_{v1}} \tag{5.41}$$

where $\Omega = \sqrt{2\boldsymbol{W} : \boldsymbol{W}}$ is the magnitude of the vorticity, and d is the distance of the flow field point to the nearest wall. $\boldsymbol{W}$ is defined by the following equation:

$$\boldsymbol{W} = \frac{1}{2} \left(\nabla \boldsymbol{v}_f - (\nabla \boldsymbol{v}_f)^{\mathrm{T}} \right) \tag{5.42}$$

$$f_{\omega} = g \left(\frac{1 + C_{\omega 3}^6}{g^6 + C_{\omega 3}^6} \right)^{\frac{1}{6}}, g = r + C_{\omega 2} \left(r^6 - r \right), r = \min \left[\frac{\tilde{\nu}}{\tilde{S} k^2 d^2}, 10 \right] \tag{5.43}$$

The model constants are:

$$C_{b1} = 0.1355, C_{b2} = 0.622, k = 0.41, \sigma = \frac{2}{3}$$

$$C_{\omega 2} = 0.3, C_{\omega 3} = 2, C_{v1} = 7.1, C_{w1} = \frac{C_{b1}}{k^2} + \frac{1 + C_{b2}}{\sigma} \tag{5.44}$$

5.3.2 *Large Eddy Simulation(LES)*

LES is a turbulence modeling technique that resolves large-scale turbulent structures in a fluid flow while modeling the smaller, subgrid scales through a turbulence model. The fundamental assumption in LES is that turbulent flows are inherently composed of structures with varying scales. Large eddies, which contribute the most to momentum and energy transport, are directly resolved, whereas the smaller eddies are modeled using a subgrid-scale (SGS) model. The governing equations for LES can be derived from the Navier–Stokes equations by filtering the velocity field at a specified length scale.

The filtered Navier–Stokes equations for LES are written as:

$$\frac{\partial \left(\alpha_f \rho_f\right)}{\partial t} + \nabla \cdot \left(\alpha_f \rho_f \overline{\boldsymbol{v}}_f\right) = 0 \tag{5.45}$$

$$\frac{\partial \left(\alpha_f \rho_f \overline{\boldsymbol{v}}_f\right)}{\partial t} + \nabla \cdot \left(\alpha_f \rho_f \overline{\boldsymbol{v}}_f \otimes \overline{\boldsymbol{v}}_f\right) = -\alpha_f \nabla \overline{p} + \nabla \cdot \alpha_f \overline{\boldsymbol{\tau}}_f + \alpha_f \rho_f \boldsymbol{g} + \overline{\boldsymbol{F}}_{pf} - \nabla \cdot \boldsymbol{\tau}^{SGS} \tag{5.46}$$

where $\overline{\boldsymbol{v}}_f$ is the filtered fluid velocity; $\overline{p}$ is the filtered pressure; $\overline{\boldsymbol{\tau}}_f$ is the filtered viscous stress tensor; $\overline{\boldsymbol{F}}_{pf}$ is the filtered particle–fluid interaction force; $\boldsymbol{\tau}^{SGS}$ is the SGS stress tensor, representing unresolved turbulence effects.

$$\boldsymbol{\tau}^{SGS} = -\rho_f \left(\overline{\boldsymbol{v}_f \otimes \boldsymbol{v}_f} - \overline{\boldsymbol{v}}_f \otimes \overline{\boldsymbol{v}}_f\right) \tag{5.47}$$

One of the major challenges in LES-DEM coupling is the computational cost. LES requires a fine grid resolution to capture large turbulent eddies, while DEM involves tracking individual particles, which can be computationally expensive, particularly for systems with a large number of particles. The need to synchronize the fluid and particulate phases at each time step adds to the complexity of the numerical solution. To address these challenges, techniques such as parallel computing, adaptive grid refinement and multiscale models are employed to optimize the efficiency of LES-DEM simulations. Another challenge is the scale disparity between the fluid and particulate phases. The time scales of the fluid dynamics, governed by turbulent eddies, are typically of the order of microseconds to milliseconds. In contrast, the response time of particles can vary widely, from microseconds for fine particles to seconds for large, massive particles. When the fluid time scale is much smaller than the particle time scale, this disparity necessitates sophisticated coupling strategies to ensure numerical stability and accuracy. Often, a time-stepping strategy is used where the fluid phase is solved on a finer time scale compared to the particulate phase, with appropriate interpolation methods to synchronize the updates between the phases.

5.3.3 Direct Numerical Simulation(DNS)

DNS is a high-fidelity computational technique used to solve the Navier–Stokes equations directly without relying on turbulence models. Unlike RANS or LES, DNS resolves all scales of turbulent motion, from the largest energy-containing eddies to the smallest dissipative Kolmogorov scales. This requires extremely fine computational grids and small time steps to accurately capture the full spectrum of turbulent fluctuations, making DNS computationally very expensive. However, it provides the most accurate representation of turbulent flows, serving as a benchmark for validating lower-fidelity models and offering deep insights into fundamental fluid dynamics.

The methodology of DNS involves discretizing the governing equation using numerical schemes such as finite difference, finite volume or spectral methods. Spectral methods are often preferred for their high accuracy in resolving small-scale turbulence, particularly in simple geometries like channel flows. The absence of turbulence closures in DNS means that all physical processes are explicitly resolved, eliminating modeling errors. However, this also imposes strict resolution requirements: the grid spacing must be smaller than the Kolmogorov length scale, and the time step must resolve the fastest fluid motions. Consequently, DNS is limited to low-to-moderate Reynolds number flows due to the prohibitive computational cost at higher Reynolds numbers.

5.3.4 Model Selection Criteria

The comparison of RANS, LES and DNS approaches is displayed in Table 5.1. Choosing the appropriate turbulence model depends on several factors, including accuracy requirements, computational resources and the nature of the flow problem. For industrial applications where computational efficiency is paramount, RANS is the preferred choice. It provides sufficiently accurate results for steady-state or time-averaged flows. However, its reliance on turbulence models means it may fail in cases with strong flow separation or highly unsteady phenomena. When transient turbulent structures significantly influence the flow, LES becomes necessary. It captures large-scale unsteadiness more accurately than RANS. DNS is reserved for fundamental research where full turbulence resolution is essential. Since it requires

Table 5.1 Comparison of RANS, LES and DNS approaches

Feature	RANS	LES	DNS
Approach	Solves time-averaged equations with turbulence models	Resolves large eddies directly and models small eddies	Solves full Navier–Stokes equations without turbulence modeling
Turbulence modeling	Uses closure models	Uses subgrid-scale models	No modeling
Computational cost	Lowest	Moderate	Extremely high
Accuracy	Low	Medium	Highest
Applicable Reynolds number (*Re*)	High-*Re* flows	Moderate-to-high *Re*	Low-to-moderate *Re*
Grid requirements	Coarse	Fine in regions of interest	Extremely fine

resolving all scales of motion, it remains impractical for high-Reynolds-number flows or large domains. However, DNS serves as a critical tool for validating RANS and LES models, providing benchmark data for turbulence physics studies. Hybrid approaches, such as Detached Eddy Simulation (DES), combine RANS and LES to balance accuracy and computational cost. DES uses RANS near walls where turbulence is more predictable and LES in separated regions where large eddies dominate, making it useful for complex aerodynamic flows.

5.4 Numerical Discretization Methods

Numerical discretization is a fundamental step in CFD that converts the continuous governing equations into a system of algebraic equations. Discretization methods are broadly classified into spatial discretization and temporal discretization. The spatial discretization deals with the approximation of derivatives across the computational domain. The temporal discretization handles the evolution of the solution over time.

5.4.1 Spatial Discretization

Discretization is the first step in numerical computation, where the continuous domain is discretized into discrete points or volumes. Common discretization methods include the finite difference method (FDM), finite volume method (FVM) and finite element method (FEM).

The FDM works by discretizing the computational domain into a grid of points and replacing the continuous differential equations with algebraic equations that approximate the derivatives at these points. FDM is particularly useful for solving problems involving partial differential equations in one or more dimensions. Its simplicity and ease of implementation make it attractive for problems with structured grids, but it struggles with complex geometries and irregular boundaries. Additionally, the accuracy and stability of FDM depend on the choice of the grid size and the time step.

The FVM is a widely used numerical approach that focuses on the conservation of quantities within small control volumes of a discretized domain. The domain is divided into a finite number of control volumes, and the integral form of the governing equations is applied to each control volume. Fluxes of quantities across the control volume boundaries are computed to ensure conservation. One of the key advantages of FVM is its ability to handle complex geometries and unstructured meshes efficiently. Unlike FDM, FVM does not directly approximate derivatives but instead focuses on flux balance. This makes FVM inherently conservative and well-suited for problems involving shocks or discontinuities, such as those in compressible fluid flow.

The FEM is a powerful and versatile numerical technique that approximates solutions to partial differential equations by dividing the domain into smaller subdomains called elements. Within each element, the solution is approximated using predefined basis functions. These elements are connected at points called nodes, and the overall solution is assembled by enforcing continuity and balance across the elements. FEM is particularly advantageous for solving problems with complex geometries, variable material properties and irregular boundaries. It is commonly used in structural analysis, heat transfer and electromagnetic simulations. FEM requires the formulation of a variational problem, typically involving minimization of an energy functional, which leads to a system of algebraic equations. Although FEM is more computationally intensive compared to FDM or FVM, its flexibility and accuracy for a wide range of applications make it one of the most popular numerical methods in engineering and physics.

5.4.2 Temporal Discretization

Temporal discretization addresses the evolution of flow variables over time. Unlike spatial derivatives, which are approximated over a computational grid, temporal derivatives require discretization across successive time steps to advance the solution from an initial state.

Explicit methods compute the solution at the next time level solely based on known quantities from the current and/or previous time levels. Explicit schemes are computationally efficient per time step since they avoid solving large systems of equations. However, they are conditionally stable, requiring the time step Δt to satisfy the Courant-Friedrichs-Lewy condition, which restricts Δt based on the grid spacing and flow velocity to prevent numerical instability. Due to this constraint, explicit methods are often employed in problems with fast transients.

Implicit methods, in contrast, incorporate unknown quantities at the new time level into the discretized equations, leading to a system of coupled algebraic equations that must be solved iteratively. The primary advantage of implicit methods is their unconditional stability, allowing for larger time steps than explicit schemes without numerical divergence. This makes them particularly suitable for stiff problems, such as diffusion-dominated flows, low-speed incompressible flows or long-time transient simulations where computational efficiency is critical.

Overall, the choice of temporal discretization depends on the physical characteristics of the flow. Explicit schemes are preferred in wave propagation and compressible flows where timescales are short and CFL constraints are manageable. Implicit schemes dominate in incompressible flows, turbulent simulations and multiphase flows with strong interphase coupling. Adaptive time stepping can dynamically adjust the time step Δt based on local flow conditions, optimizing both stability and efficiency.

5.5 Geometry Modeling and Mesh Generation

5.5.1 Geometry Modeling

Geometry modeling is the process of creating the physical domain in which the fluid flow or thermal phenomena are to be simulated. This step is essential as it determines the boundaries, the features, and the structure of the domain where the fluid will interact. Accurate geometry modeling is crucial because any error or simplification at this stage can lead to inaccurate simulation results. The complexity of the geometry can vary widely depending on the problem. The geometry could be two-dimensional, three-dimensional or axisymmetric. Several software tools are available for geometry modeling, such as AutoCAD, SolidWorks, CATIA and ANSYS Design Modeler. These tools allow engineers and researchers to construct the geometries with high precision, incorporating intricate details such as curves, filets, holes and other features that are necessary for the simulation. After the geometry is modeled, it often needs to be cleaned up to remove any unnecessary features, such as small edges or internal faces that could complicate the meshing process. Geometry simplification is common in industrial CFD applications, especially when dealing with complex shapes, as simplifying the geometry can reduce computational costs without significantly affecting the accuracy of the simulation.

5.5.2 *Mesh Generation*

Once the geometry is defined, the next step in CFD is mesh generation, which involves discretizing the physical domain into smaller elements or cells that can be used to approximate the solution of the governing equations. The quality of the mesh directly affects the accuracy of the results. A good mesh should accurately capture the fluid flow characteristics and the geometry while minimizing computational resources. There are several types of meshes that can be used in CFD. In a structured mesh, the grid points are arranged in a regular pattern, such as a square grid in 2D or a cubic grid in 3D. This type of mesh is easy to generate and provides high accuracy, especially for problems with regular geometries. However, structured meshes become impractical for complex, irregular geometries. In an unstructured mesh, the grid points are distributed irregularly, allowing for more flexibility in meshing complex geometries. These meshes use polygons or polyhedra, such as triangles or tetrahedra, for 2D and 3D problems, respectively. Although unstructured meshes are more computationally expensive, they are the preferred choice for complicated geometries or when a high level of accuracy is required in localized regions of the domain. Hybrid meshes combine structured and unstructured elements in the same domain. This approach is often used in CFD simulations where structured meshes are applied in areas of regular geometry, and unstructured meshes are used in regions where the geometry becomes more complex.

Mesh quality plays an important role in the accuracy and convergence of CFD simulations. Poor-quality meshes, characterized by high aspect ratios, skewed elements or large variations in cell sizes, can lead to inaccurate results and slow convergence. The aspect ratio of each cell should be close to 1, meaning the cells should be as close to squares or cubes as possible. Extreme aspect ratios can lead to numerical instability. Skewed cells can distort the results. Ideally, cells should be as orthogonal as possible to the boundaries. The mesh should be refined in areas of high gradient such as boundary layers or regions with steep gradients in velocity or temperature and coarse in areas where the flow is more uniform. Adaptive mesh refinement techniques can automatically refine the mesh where needed during the simulation. In practice, mesh generation is an iterative process. After an initial mesh is created, it is tested by running preliminary simulations. The results are analyzed to determine whether the mesh resolution is adequate or whether further refinement is necessary.

5.6 Boundary Conditions

Boundary conditions define the interaction of the fluid with the physical boundaries of the computational domain. These conditions specify the values of the flow variables such as velocity, pressure, temperature at the boundaries. The choice of boundary conditions can significantly impact the accuracy and reliability of CFD results. There are several types of boundary conditions, each representing different physical situations.

(1) Inlet boundary conditions

Inlet boundary conditions define how fluid enters the computational domain. The most common types include velocity inlet, pressure inlet and mass flow inlet. Velocity inlet specifies a fixed velocity profile at the inlet, often used for incompressible flows. The pressure is then extrapolated from the interior domain. This condition is suitable when the inflow velocity is known, such as in pipe flows or wind tunnel simulations. Pressure inlet prescribes total pressure and static pressure, typically used in compressible flows or cases where the inlet pressure is known but velocity may vary. Mass flow inlet directly imposes a mass flow rate rather than velocity, useful in systems where mass conservation is critical such as in combustion chambers.

(2) Outlet boundary conditions

Outlet conditions define how flow exits the domain. The two most widely used types are pressure outlet and outflow. Pressure outlet specifies static pressure at the outlet, allowing velocity to adjust naturally. This is common in external aerodynamics and internal flows where the exit pressure is known. If backflow occurs, additional specifications such as backflow turbulence intensity may be required. Outflow assumes fully developed flow, where all flow variables except pressure have a zero normal gradient.

(3) Wall boundary conditions

Wall boundary conditions represent perhaps the most physically significant constraints in many CFD simulations, as they define the interaction between the fluid and solid surfaces. The no-slip condition, which enforces zero relative velocity between the fluid and wall, is fundamental for viscous flows and essential for accurate resolution of boundary layers in turbulent flow simulations. In certain specialized applications, such as high-altitude aerodynamics or inviscid flow modeling, a slip condition may be employed instead, allowing for tangential flow movement without frictional effects. Moving wall conditions extend this concept to rotating or translating surfaces, which are particularly important in turbomachinery applications and vehicle aerodynamics. For rough surfaces, specialized wall functions incorporate surface roughness effects into turbulence modeling, modifying the near-wall flow behavior to match empirical observations. The treatment of wall boundaries must be carefully coordinated with the chosen turbulence model and near-wall mesh resolution to ensure physically accurate results.

(4) Symmetry and periodic boundary conditions

Symmetry and periodic boundary conditions offer valuable computational efficiencies by reducing the required domain size through exploitation of flow field repetitions or symmetrical properties. Symmetry conditions, which enforce zero normal velocity and zero shear stress at symmetry planes, are commonly used in simulations of axisymmetric or geometrically symmetric flows. Periodic conditions take this concept further by assuming spatial periodicity in the flow field, enabling the simulation of only a single repeating unit in applications like heat exchanger banks or turbine blade cascades. However, these conditions must be applied with caution as incorrect assumptions of symmetry or periodicity in inherently asymmetric or aperiodic flows can introduce significant errors into the simulation results.

5.7 Near Wall Treatment

The near-wall region is where the most significant fluid velocity gradients occur due to the no-slip condition, which states that the fluid velocity at the boundary is zero. This creates a boundary layer where the velocity gradually increases from zero at the wall to the free-stream velocity further away from the surface. In turbulent flows, the near-wall region also involves complex interactions between large-scale turbulent eddies and the wall itself, which leads to highly variable velocity profiles and the potential for phenomena such as flow separation and recirculation. Due to the complexity of the near-wall region, different methods and models are used in CFD to account for the different flow regimes and turbulence behaviors near boundaries.

In turbulent flow near walls, the region can be divided into several layers, each characterized by different flow dynamics. The viscous sublayer is the region closest to the wall, where the flow is dominated by viscous forces, and the turbulence is largely suppressed. In the viscous sublayer, the flow is almost laminar, and the velocity profile follows a linear relationship with distance from the wall. The velocity gradient is very steep in this layer, and the flow behaves similarly to a viscous, steady state. In turbulent flow, the Reynolds number is very low in this region, and the turbulence intensity is negligible. Above the viscous sublayer, the buffer layer acts as a transition zone where both viscous and turbulent forces coexist. This layer connects the viscous sublayer to the outer turbulent boundary layer. Here, the velocity gradient decreases as the influence of turbulence starts to grow. The flow is neither fully laminar nor fully turbulent in this region, and turbulence is still in its developing stage. The turbulent boundary layer is the outermost layer near the wall where turbulence is fully developed. In this region, the flow is dominated by large-scale turbulent eddies, and the velocity profile follows a logarithmic relationship with distance from the wall. The turbulent shear stress is significant, and it contributes greatly to the overall momentum transfer in the fluid. The velocity profile in this region is less sensitive to viscosity and more controlled by the turbulence structure.

The key challenge in CFD simulations is the accurate representation of these different layers in the near-wall region, while maintaining computational efficiency. Several approaches are commonly used for near-wall treatment, depending on the resolution of the mesh, the turbulence model employed and the flow characteristics. Wall functions are an approximation used in CFD to avoid resolving the fine details of the near-wall region especially the viscous sublayer by adjusting the turbulence model in this region. Wall functions allow for a coarser mesh near the wall, making the simulation more computationally efficient. Empirical relationships are used based on the log-law to predict the behavior of the flow in the log-layer and the buffer region. Wall functions are particularly useful in simulations where the mesh is not fine enough to resolve the entire boundary layer. For more accurate near-wall treatment, particularly in highly viscous flows or where detailed turbulence structures are important, low-Reynolds number turbulence models can be used. These models attempt to resolve the flow in the near-wall region by using a finer mesh and accounting for the effects of viscosity across the entire boundary layer, including the viscous sublayer. Low-Reynolds models modify standard turbulence models such as the k-ε or k-ω models to incorporate the effects of viscosity near the wall. They are able to directly resolve the turbulence characteristics in the viscous sublayer and buffer layer, which improves the accuracy of predictions for wall shear stress, heat transfer and turbulence intensity. However, these models require a much finer mesh near the wall, which increases computational cost. They also tend to be more complex and may require more detailed flow information to be implemented properly. Enhanced wall treatment is a hybrid approach that combines the benefits of wall functions and low-Reynolds number models. This method allows to automatically switch between wall functions for coarse meshes and low-Reynolds number models for fine meshes. The enhanced wall treatment can resolve the viscous sublayer when the mesh is fine enough, while still providing a reasonable approximation with wall functions when the mesh resolution is coarser. It provides a good balance between accuracy and computational efficiency.

References

1. L. Schiller, Z. Naumann, Über die grundlegenden Berechnungen bei der Schwerkraftaufbereitung. Ingenieure **77**, 318–320 (1933)
2. C.Y. Wen, Y.H. Yu, A generalized method for predicting the minimum fluidization velocity. AIChE J. **12**(3), 610–612 (1966). https://doi.org/10.1002/aic.690120343
3. D. Gidaspow, Multiphase flow and fluidization: continuum and kinetic theory descriptions. AIChE J. **42**(4), 1197–1198 (1994). https://doi.org/10.1002/aic.690420438
4. M. Syamlal, T.J. O'Brien, Computer simulation of bubbles in a fluidized bed, in *AIChE Symp. Ser*, Publ by AIChE (1989), pp. 22–31
5. Y. Igci, A.T. Andrews IV., S. Sundaresan, S. Pannala, T. O'Brien, Filtered two-fluid models for fluidized gas-particle suspensions. AIChE J. **54**(6), 1431–1448 (2008). https://doi.org/10.1002/aic.11481
6. J. Li, M. Kwauk, *Particle-fluid two-phase flow: the energy-minimization multi-scale method* (Metallurgical Industry Press, Beijing, 1994)
7. K. Agrawal, P.N. Loezos, M. Syamlal, S. Sundaresan, The role of meso-scale structures in rapid gas–solid flows. J. Fluid Mech. **445**, 151–185 (2001). https://doi.org/10.1017/S0022112001005663
8. J.F. Parmentier, O. Simonin, O. Delsart, A functional subgrid drift velocity model for filtered drag prediction in dense fluidized bed. AIChE J. **58**(4), 1084–1098 (2012). https://doi.org/10.1002/aic.12647
9. ANSYS Inc (Release-15.0), *ANSYS Fluent Theory Guide*, vol. 15317 (ANSYS Inc., USA, vol. 15317, 2015), pp. 1–759

10. H. Xu, T.J. Wray, R.K. Agarwal, Application of a new DES model based on Wray-Agarwal turbulence model for simulation of wall-bounded flows with separation, in *47th AIAA Fluid Dynamics Conference* (2017), pp. 1–23. https://doi.org/10.2514/6.2017-3966
11. P. Spalart, S. Allmaras, A one-equation turbulence model for aerodynamic flows, in *30th Aerospace Sciences Meeting and Exhibit* (American Institute of Aeronautics and Astronautics, 1992). https://doi.org/10.2514/6.1992-439

Chapter 6
Fluid-Particle Coupling Schemes

6.1 Coupling Resolution Frameworks

In fluid-particle coupling schemes, the resolution frameworks are categorized based on the ratio of the cell size ΔL to the particle diameter d_p. These frameworks determine how the interaction between fluid and particles is resolved and include unresolved, semi-resolved and resolved CFD-DEM coupling methods, as shown in Fig. 6.1.

6.1.1 Unresolved CFD-DEM Coupling

The unresolved CFD-DEM coupling method is the most widely used approach for simulating large-scale particle–fluid systems due to its computational efficiency. In this framework, the size of the CFD cells is much larger than the particle diameter, typically with a ratio of $\Delta L/d_p \geq 3$, meaning that individual particle boundaries are not explicitly resolved. Instead, the fluid phase is generally solved using locally averaged Navier–Stokes equations, where the presence of particles is accounted for through the voidage fraction α_f, representing the fluid volume fraction in each cell. A key aspect of unresolved coupling is the empirical modeling of particle–fluid interaction forces, particularly the drag force, which dominates momentum exchange. Popular drag models, such as those by Gidaspow [2] and Wen and Yu [3], are employed, incorporating the voidage fraction and relative velocity between phases. However, since the particle boundaries are not resolved, these models rely on assumptions that may introduce errors, especially for non-spherical particles or systems with strong local flow gradients.

To compute the voidage fraction, the particle centroid method (PCM) is commonly used, where each particle's entire volume is assigned to the CFD cell containing its centroid. While simple, PCM can lead to abrupt changes in α_f as particles cross cell

R. K. Agarwal and Y. Shao, *Theory and Application of Discrete Element Method to Simulation of Multiphase Flows*, https://doi.org/10.1007/978-3-032-11157-9_6

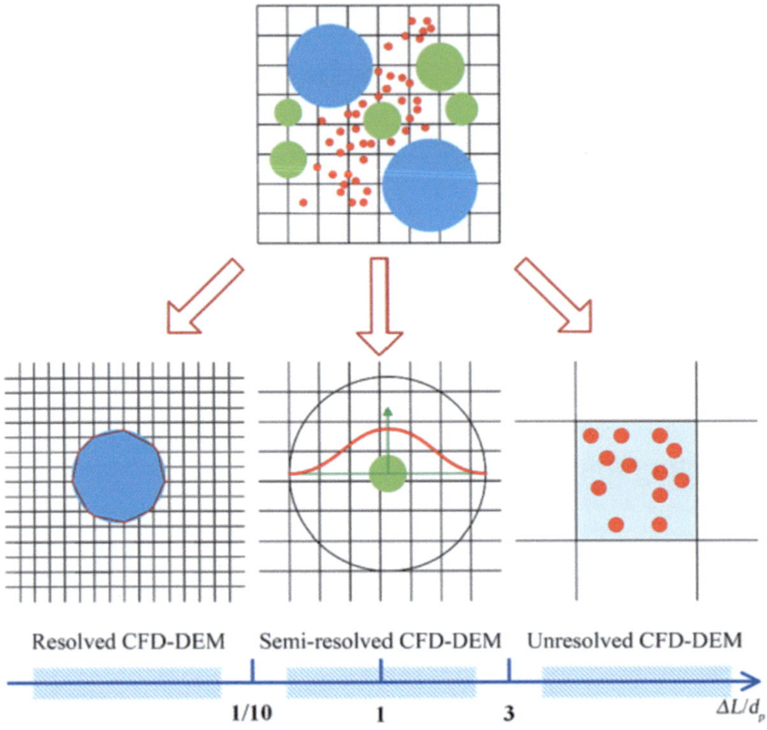

Fig. 6.1 Schematic illustration of the resolved CFD-DEM, semi-resolved CFD-DEM and unresolved CFD-DEM [1]

boundaries, causing numerical instability. Improved methods, such as the divided particle volume approach or satellite point method, distribute particle volumes across neighboring cells, ensuring smoother transitions and better accuracy. Figure 6.2 displays the sketch map of calculating the voidage fraction in the two methods.

The unresolved approach is highly efficient and suitable for simulating dense particle systems, such as fluidized beds, pneumatic conveying and sedimentation

Fig. 6.2 Sketch map of calculating the voidage fraction: **a** PCM and **b** the improved method [4]

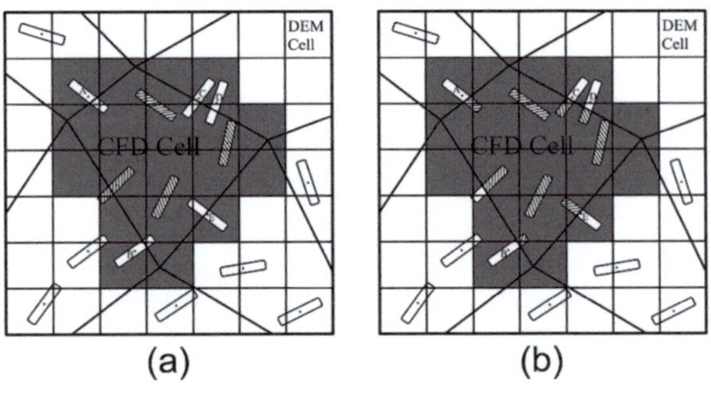

processes, where resolving individual particle–fluid interactions is computationally prohibitive. However, its reliance on empirical drag correlations limits accuracy in systems where microscale flow details such as boundary layers and particle wakes significantly influence overall behavior. Despite this, unresolved CFD-DEM remains the preferred choice for industrial-scale simulations due to its balance between computational cost and practical predictive capability.

6.1.2 Semi-resolved CFD-DEM Coupling

The semi-resolved CFD-DEM coupling method improves accuracy over unresolved methods while maintaining reasonable computational efficiency. In this method, the fluid mesh is refined to a scale where individual particles begin to influence the local flow field, typically with a cell size ranging from 0.1 to 3 particle diameters. This intermediate resolution allows for better representation of particle–fluid interactions compared to fully unresolved methods, while avoiding the excessive computational costs of fully resolved simulations. The method achieves this balance through several key features that distinguish it from other coupling approaches. A defining characteristic of semi-resolved coupling is its treatment of the void fraction field. Unlike unresolved methods that simply assign particle volumes to containing cells, semi-resolved approaches employ sophisticated interpolation techniques to distribute particle influences across neighboring fluid cells. Common implementations include kernel-based smoothing methods that weight the particle's effect on surrounding fluid based on distance, or more advanced techniques like the diffused interface method that solves additional transport equations to smooth the phase boundaries. These approaches significantly reduce the numerical artifacts that plague traditional unresolved methods when particles move across cell boundaries.

The calculation of fluid-particle interaction forces in semi-resolved coupling benefits from the improved spatial resolution. While still relying on empirical correlations for the overall force magnitude, the method can better account for local flow conditions around each particle. Some implementations incorporate analytical solutions for simple flows around canonical particle shapes, blending these with the background numerical solution to capture near-particle flow features that would be missed in unresolved simulations. This proves particularly valuable for non-spherical particles, where orientation-dependent effects on drag and lift can be partially resolved.

For industrial-scale simulations involving complex particle shapes, semi-resolved coupling offers practical advantages. The method can handle thousands to millions of particles while still capturing important phenomena like particle clustering, preferential concentration in turbulent flows and orientation-dependent settling behavior. Its moderate computational requirements make it feasible for parametric studies and optimization tasks where numerous simulations are required. However, the approach does face certain limitations. The accuracy of force calculations still depends on the quality of underlying empirical models, and the method cannot fully resolve boundary layer effects or lubrication forces between nearly touching particles. The optimal choice of smoothing parameters or interpolation schemes may require case-specific tuning.

6.1.3 *Resolved CFD-DEM Coupling*

Resolved CFD-DEM coupling represents the most accurate but computationally demanding approach for simulating particle–fluid interactions. In this method, the fluid mesh is refined to such an extent (typically $\Delta L/d_p \leq 0.1$) that the flow around each individual particle is fully resolved, eliminating the need for empirical drag correlations. This approach directly solves the Navier–Stokes equations around particle boundaries, capturing all relevant fluid dynamics phenomena including boundary layers, wake formations and hydrodynamic interactions between nearby particles.

The key advantage of resolved coupling lies in its first-principles treatment of fluid-particle interactions. Unlike unresolved or semi-resolved methods that rely on approximate drag laws, resolved coupling computes hydrodynamic forces and torques directly from the stress integration over particle surfaces. This makes it particularly valuable for studying non-spherical particles, where orientation-dependent drag and lift forces can be captured naturally. The method typically employs either body-fitted meshes that conform to particle surfaces or immersed boundary methods (IBMs) that represent particles through forcing terms in the fluid equations.

For industrial applications, resolved CFD-DEM faces significant computational challenges. The need to resolve thin boundary layers around particles often requires extremely fine meshes, limiting practical simulations to systems containing at most a few thousand particles. Recent advances in adaptive mesh refinement and GPU-accelerated computing have helped extend these limits, but resolved simulations remain orders of magnitude more expensive than their unresolved counterparts. The method finds its niche in fundamental studies of particle–fluid interactions, development of improved drag models for coarser methods and validation of industrial-scale simulations in representative small-scale systems.

6.2 Coupling Directionality

In CFD-DEM simulations, the coupling directionality defines how particles and fluid interact, ranging from simple one-way coupling to complex four-way coupling. The choice depends on the particle concentration, flow regime and required accuracy.

6.2.1 One-Way Coupling

One-way coupling represents the most basic interaction scheme in coupled CFD-DEM simulations, where the fluid phase influences particle motion without any reciprocal feedback from particles to the fluid. This unidirectional approach is typically implemented in unresolved or semi-resolved coupling frameworks, as it inherently assumes that particle presence does not modify the underlying flow field, an assumption that aligns with these coarser resolution methods.

In the unresolved framework, where the CFD cell size is much larger than the particle diameter, one-way coupling simply extracts fluid velocities from the pre-solved flow field at particle positions to compute hydrodynamic forces. The fluid equations are solved independently of particles, making this approach computationally efficient but limited to extremely dilute systems. The lack of voidage feedback means traditional unresolved methods like the PCM can be used without modification, though interpolation schemes may still be applied for smoother force calculations. For slightly more refined scenarios, one-way coupling can also operate in a semi-resolved framework while particles still don't affect the fluid, the improved spatial resolution allows for better estimation of local fluid velocities around non-spherical particles or in shear flows. This is particularly useful when studying particle orientation dynamics in known flow fields. Overall, the key advantage of one-way coupling lies in its simplicity. However, this comes with strict limitations: it cannot capture particle-induced turbulence, clustering effects or any flow modifications from particle momentum exchange.

6.2.2 Two-Way Coupling

Two-way coupling represents a more complex and physically accurate approach compared to one-way coupling. In this scheme, both the fluid and the particle phases interact with each other, meaning that not only do the fluid forces influence the particles, but the particles also exert forces back onto the fluid. This mutual interaction is crucial for accurately modeling systems where particles significantly affect the fluid flow and vice versa. Two-way coupling is typically applied in scenarios involving dense particle-laden flows, such as in fluidized beds, slurry transport and particle suspension systems, where both the particles and the fluid phase play active roles in shaping the behavior of the system.

The implementation of two-way coupling requires careful consideration of several key aspects. First, the fluid phase equations must be modified to include source terms representing momentum exchange from particles to fluid. These source terms are calculated by summing the reaction forces from all particles in each CFD cell and distributing them appropriately across the fluid mesh. Second, accurate calculation of the local void fraction becomes crucial, as it directly affects both the modified Navier–Stokes equations and the interphase momentum transfer terms. The two-way coupling approach naturally accommodates different resolution frameworks. In unresolved simulations, it relies heavily on empirical drag correlations to calculate interphase forces, while in semi-resolved cases, it can partially capture near-particle flow features while still using corrected drag models. The choice of resolution framework significantly impacts both accuracy and computational cost.

A critical advantage of two-way coupling is its ability to capture important physical phenomena that one-way coupling cannot, such as particle-induced turbulence, bubble formation in fluidized beds, and the collective behavior of particle clusters in pneumatic conveying. However, it still has limitations, particularly in dense systems where particle–particle interactions mediated by the fluid become important. These limitations have led to the development of more sophisticated four-way coupling approaches for high-concentration systems.

6.2.3 Four-Way Coupling

Four-way coupling represents the most comprehensive approach for simulating dense particle–fluid systems, extending beyond two-way coupling by explicitly accounting for both particle–fluid and fluid-mediated particle–particle interactions. This advanced coupling scheme becomes essential when particle concentrations exceed 10–20% volume fraction, where short-range hydrodynamic forces between nearby particles significantly influence system behavior. Unlike two-way coupling that only considers mean fluid-particle interactions, four-way coupling resolves the intricate fluid dynamics in the narrow gaps between particles, enabling accurate prediction of phenomena like hydrodynamic clustering.

The implementation of four-way coupling typically requires a resolved or highly refined semi-resolved framework to properly capture the fluid flow in interparticle regions. In this approach, the fluid phase not only exchanges momentum with individual particles but also transmits forces between particles through the intervening fluid. This is particularly important for viscous systems where lubrication forces dominate particle interactions at small separations. The method often employs specialized techniques such as IBMs with adaptive mesh refinement near-particle surfaces or lattice Boltzmann methods that naturally handle complex boundary conditions. These computational strategies allow for direct resolution of the thin fluid films between particles that would otherwise be averaged out in coarser approaches.

Four-way coupling is computationally intensive, as it requires the solution of both fluid dynamics and particle mechanics simultaneously, with significant cross-coupling between the phases. Additionally, high-fidelity models are often required to handle the complex interactions between particle phases. This increases the computational burden substantially compared to two-way and one-way coupling. However, the increased accuracy and predictive power provided by four-way coupling make it indispensable for applications in areas such as chemical reactors, mineral processing, pharmaceutical formulations and aerosol dynamics, where the interactions between different phases and particle types are crucial for understanding process dynamics and optimizing design.

6.3 Numerical Implementation Methods

The numerical implementation of fluid-particle coupling requires careful selection of methods that align with both the coupling resolution framework and the coupling directionality. These choices describe how fluid-particle interactions are mathematically represented and computationally solved. Two common implementation methods include averaged volume method (AVM) and IBM.

6.3.1 Averaged Volume Method (AVM)

In AVM, the fluid grid resolution is intentionally coarser than individual particle sizes. This method is particularly valuable in industrial-scale simulations involving large numbers of particles, as it circumvents the computational expense of resolving each particle's boundary layer by employing a locally averaged formulation of the Navier–Stokes equations. The fundamental premise of AVM lies in treating the influence of particles on the fluid phase through continuum approximations, incorporating its presence through two key modifications to the fluid equations: the introduction of a void fraction field to account for volume displacement by particles, and the addition of momentum exchange terms to represent fluid-particle interactions. These modifications enable efficient simulation of systems where microscale flow features around individual particles are less critical than the bulk behavior of the particle–fluid mixture.

At the heart of AVM's implementation is the calculation of void fraction, which quantifies the fluid volume fraction in each computational cell. This is typically achieved by distributing the volume of each particle to nearby fluid grid points using weighting functions, ranging from simple particle centroid-based methods to more sophisticated Gaussian kernels. The method then computes fluid-particle interaction forces, predominantly drag forces, using well-established empirical correlations such as the Wen-Yu model for dilute flows or the Ergun equation for dense regimes. These correlations relate the relative velocity between fluid and particles to the drag force through empirically determined coefficients, effectively bridging the scale gap between the unresolved particles and the resolved fluid flow. The numerical coupling follows a sequential approach where fluid velocities are interpolated to particle positions, forces are calculated, and then projected back to the fluid grid in a conservative manner, creating a two-way coupled system that captures the essential physics of fluid-particle interaction while maintaining computational tractability.

While AVM offers significant computational advantages, it comes with inherent limitations tied to its reliance on empirical closures and averaging assumptions. The method's accuracy is contingent upon the validity of the drag correlations employed, which may break down for non-spherical particles, wide particle size distributions or in flow regimes far from those in which the correlations were developed. Furthermore, the averaging process smoothes out microscale phenomena such as particle wakes and near-field hydrodynamic interactions, which could be critical in certain applications.

6.3.2 Immersed Boundary Method (IBM)

IBM represents a sophisticated class of numerical techniques for resolving fluid-particle interactions at the individual particle scale, offering a critical advantage over averaged approaches when detailed flow features around particles must be captured. These methods are particularly valuable in resolved and semi-resolved CFD-DEM coupling scenarios, where the fluid grid is sufficiently refined to explicitly represent particle boundaries. The fundamental principle involves embedding Lagrangian markers for rigid bodies or meshes for deformable bodies in a fixed Eulerian fluid grid, allowing for precise enforcement of no-slip boundary conditions without the computational burden of dynamic mesh regeneration. This approach stands in contrast to the AVM by preserving discrete particle identities while accurately resolving the surrounding flow field.

The classic IBM is the most widely used variant. At each time step, fluid velocities are interpolated from the Eulerian grid to discrete Lagrangian markers representing particle surfaces, where correction forces are computed to satisfy the no-slip condition. These forces are then projected back to the fluid grid through carefully designed spreading operators, typically implemented using smoothed delta functions that maintain numerical stability while conserving momentum. For more complex scenarios involving deformable particles, the immersed finite element method extends this framework by coupling the fluid solver with a finite element representation of the solid phase, enabling simulation of flexible structures. This extended capability allows the IBM framework to handle both rigid and deformable particles within the same computational framework, significantly expanding its range of applicability.

One of the key advantages of immersed methods is that they allow for arbitrary motion of the solid objects without the need to modify the fluid grid, making them particularly suitable for simulating dynamic and complex systems. For example, in fluid-particle interactions, where particles may rotate, move or even collide, immersed methods provide a flexible and efficient framework for handling such motion while maintaining the computational efficiency of the underlying CFD solver. This makes immersed methods particularly useful in applications involving particle-laden flows, where particle dynamics play a significant role in determining the overall system behavior. However, these advantages come with significant computational demands, as resolving individual particle boundaries requires fine grids and small time steps that can limit system size and simulation duration.

6.4 Computational Sequence

The computational sequence in a CFD-DEM coupling typically involves a series of iterative steps designed to solve the governing equations for both the fluid and solid phases while accounting for their mutual interactions.

The computation initiates with a comprehensive initialization phase, where the fluid domain is discretized using Eulerian grids with appropriate boundary conditions including inlet/outlet/wall specifications. The discrete phase is initialized by defining particle positions, velocities and material properties such as density, diameter and Young's modulus, along with contact parameters for collision modeling. Critical coupling parameters are established at this stage, including time step sizes constrained by both fluid conditions and particle contact dynamics, as well as convergence tolerances for interfacial force equilibrium.

During each time step, the solution proceeds through three principal operations. First, the fluid phase equations are solved using either volume-averaged Navier–Stokes formulations with empirical closure models for unresolved coupling, or immersed boundary techniques for resolved cases. The former employs locally averaged quantities and momentum exchange terms, while the latter explicitly resolves particle boundaries through distributed Lagrange multipliers. Concurrently, the DEM advances particle states by integrating Newton equations, incorporating hydrodynamic forces such as drag force and lift force. For four-way coupled systems, interparticle interactions are modeled through rigorous contact mechanics formulations.

The interfacial coupling is enforced through sophisticated data transfer operations. Fluid phase quantities like velocity are mapped to particle locations through appropriate interpolation operators like kernel-based methods, while particle-induced forces are projected onto fluid grid nodes using momentum-conserving schemes. This exchange is typically iterated within each time step until achieving force convergence, with the number of sub-iterations adjusted dynamically based on residual behavior. Modern implementations often employ asynchronous parallelization, allowing CFD and DEM solvers to execute concurrently on distributed computing resources while synchronizing during data exchange phases. The sequence concludes with output generation at prescribed intervals, recording coupled field data such as fluid velocity/pressure and particle positions/forces in formats suitable for both quantitative analysis and visualization.

References

1. H. Ma, L. Zhou, Z. Liu, M. Chen, X. Xia, Y. Zhao, A review of recent development for the CFD-DEM investigations of non-spherical particles. Powder Technol. **412**, 117972 (2022). https://doi.org/10.1016/j.powtec.2022.117972
2. D. Gidaspow, Multiphase flow and fluidization: continuum and kinetic theory descriptions. AIChE J. **42**(4), 1197–1198 (1994). https://doi.org/10.1002/aic.690420438

3. C.Y. Wen, Y.H. Yu, A generalized method for predicting the minimum fluidization velocity. AIChE J. **12**(3), 610–612 (1966). https://doi.org/10.1002/aic.690120343
4. H. Ma, Y. Zhao, Y. Cheng, CFD-DEM modeling of rod-like particles in a fluidized bed with complex geometry. Powder Technol. **344**, 673–683 (2019). https://doi.org/10.1016/j.powtec.2018.12.066

Chapter 7
Heat and Mass Transfer in CFD-DEM

7.1 Fundamentals of Heat Transfer

7.1.1 Governing Equations of Heat Transfer

Heat transfer in CFD-DEM simulations involves energy exchange between the fluid phase and the solid phase, as well as within each phase. The governing equations of heat transfer are derived from the conservation of energy and are solved for both phases.

For spherical particles, the lumped capacitance method is often used when the Biot number (B_i) is small ($B_i < 0.1$), meaning the temperature gradient within a particle is negligible [1], as shown in Fig. 7.1. The energy balance for particle i is given in [2]:

$$m_i c_{p,i} \frac{\mathrm{d}T_i}{\mathrm{d}t} = \sum_{j=1}^{n_i} q_{ij} + q_{i,\mathrm{wall}} + q_{if} + q_{i,\mathrm{rad}} + q_{rxn} + S_q \tag{7.1}$$

where m_i is the mass of particle i; $c_{p,i}$ is the specific heat capacity of particle i; T_i is the temperature of particle i; all q terms represent heat transfer rate; q_{ij} is the conductive heat transfer between particles i and j; $q_{i,\mathrm{wall}}$ is the conductive heat transfer between particle i and the wall; q_{if} is the convective heat transfer between particle i and the surrounding fluid; $q_{i,\mathrm{rad}}$ is the radiative heat transfer from particle i; q_{rxn} is the heat generation/consumption rate due to chemical reactions; S_q is the source term for other heat mechanism.

For conditions with $B_i > 0.1$ (large particles or high conductivity materials), the temperature distribution within a particle must be considered by solving [3]:

$$\rho_p c_p \frac{\partial T_i}{\partial t} = \frac{1}{r^2} \frac{\partial}{\partial r} \left(k_p r^2 \frac{\partial T_i}{\partial r} \right) \tag{7.2}$$

R. K. Agarwal and Y. Shao, *Theory and Application of Discrete Element Method to Simulation of Multiphase Flows*, https://doi.org/10.1007/978-3-032-11157-9_7

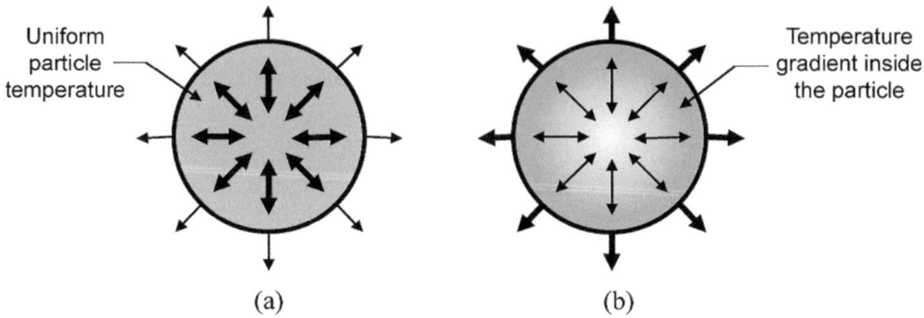

Fig. 7.1 Effects of Biot number on the particle temperature distribution: **a** small value; **b** large value [1]

where k_p is the thermal conductivity of a particle. This equation requires an initial temperature profile and boundary conditions.

The energy equation for the fluid phase is derived from the volume-averaged Navier–Stokes equations and is solved for the local average fluid temperature T:

$$\frac{\partial}{\partial t}\left(\alpha_f \rho_f c_{pf} T\right) + \nabla \cdot \left(\alpha_f \rho_f \mathbf{v}_f c_{pf} T\right) = \nabla \cdot \left(\alpha_f k_f^{\text{eff}} \nabla T\right) + Q_{fp} + Q_{f,\text{wall}} + Q_{f,rxn} \tag{7.3}$$

where α_f is the void fraction; ρ_f is the fluid density; c_{pf} is the fluid specific heat capacity; $\mathbf{v}_f$ is the fluid velocity; k_f^{eff} is the effective thermal conductivity of the fluid phase, which may account for molecular diffusion, turbulence and dispersion effects; Q_{fp} is the heat exchange between fluid and particles per unit volume, calculated as the sum of all convective heat transfer rates from particles within a fluid cell divided by the cell volume; $Q_{f,\text{wall}}$ is the heat exchange between fluid and walls; $Q_{f,rxn}$ is the heat generation/consumption due to homogeneous reactions.

7.1.2 Heat Transfer Mechanisms

The three main types of heat transfer in CFD-DEM are conduction, convection and radiation. Each of these mechanisms operates based on different physical principles and plays a role depending on the nature of the particles and the surrounding environment. Specifically, the heat transfer might involve the following mechanisms [4]:

- conduction through the fluid in the gap between two non-contacting particles;
- conduction through the contact area between two area-contacting particles;
- convection between fluid and solid particles;
- convection between fluid and wall;
- radiation between the surfaces of solid particles;

- radiation between fluid and solid particles.

(1) Conduction

Figure 7.2 displays the relative positions of two particles. Particle–particle conductive heat transfer occurs through two primary mechanisms: particle–fluid-particle conduction and direct particle–particle conduction.

$$q_{ij} = q_{ij}^{p-f-p} + q_{ij}^{p-p} \tag{7.4}$$

Particle–fluid-particle interaction occurs when particles are near but not in direct contact, with heat transferring through the intervening fluid [5]:

$$q_{ij}^{p-f-p} = \left(T_j - T_i\right) \int_{r_{ij}}^{r_{pf}} \frac{2\pi\, r dr}{\left(\sqrt{R^2 - r^2} - \frac{r(R+H)}{r_{ij}}\right) \cdot \left(\frac{1}{k_i} + \frac{1}{k_j}\right) + \frac{2\left[(R+H) - \sqrt{R^2 - r^2}\right]}{k_f}} \tag{7.5}$$

where k_i, k_j and k_f are the thermal conductivities of particle i, particle j and fluid, respectively; H is the surface-to-surface distance between particles; r_{ij} is the radius of the lens of fluid between particles.

$$H = \frac{d_{ij} - 2R}{2} \tag{7.6}$$

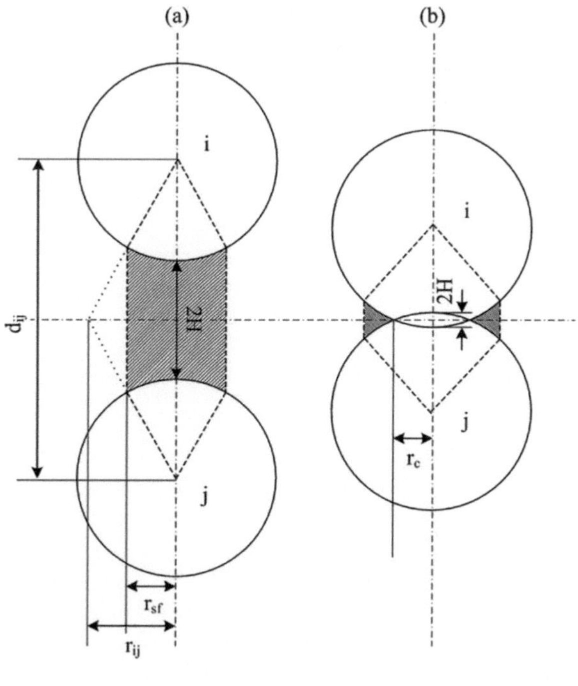

Fig. 7.2 Relative positions of two particles: **a** not in contact and **b** in contact [5]

$$r_{sf} = \frac{Rr_{ij}}{\sqrt{r_{ij}^2 + (R+H)^2}} \tag{7.7}$$

Direct conduction occurs when two particles are in physical contact, allowing heat transfer through the contact points. The heat transfer rate is driven by the temperature gradient between the particles and the thermal properties of the materials involved. The direct particle–particle conduction occurs through two sub-mechanisms: static contact conductions q_{ij}^{static} and collisional conduction $q_{ij}^{\text{collision}}$.

For particles in prolonged contact, the static contact conductions q_{ij}^{static} can be calculated as [6]:

$$q_{ij}^{\text{static}} = \frac{4r_c\left(T_j - T_i\right)}{\frac{1}{k_i} + \frac{1}{k_j}} \tag{7.8}$$

where r_c is the contact radius during collision.

For heat transfer during brief collisions, the collisional conduction $q_{ij}^{\text{collision}}$ is expressed as [7]:

$$q_{ij}^{\text{collision}} = \frac{C'\left(T_j - T_i\right)\pi r_c^2 t_c^{-\frac{1}{2}}}{(\rho_i c_i k_i)^{-\frac{1}{2}} + \left(\rho_j c_j k_j\right)^{-\frac{1}{2}}} \tag{7.9}$$

where t_c is the contact duration; C' is the empirical coefficient which is a function of Fourier number.

As for the particle–wall conduction, equations similar to the particle–particle conduction are used, treating the wall as a particle with infinite radius and mass. The contact area and duration calculations are adjusted accordingly.

It is worth mentioning that in DEM simulations, Young's modulus of particles is often artificially reduced to allow larger time steps and improve computational efficiency. However, this reduction affects the calculated contact area between particles, which directly influences conductive heat transfer. The actual contact radius $r_{c,0}$ is related to the reduced Young's modulus used in simulations by [8]:

$$r_{c,0} = r_c\left(\frac{E_{\text{actual}}}{E_{\text{sim}}}\right)^{\frac{1}{5}} \tag{7.10}$$

where r_c is the contact radius computed in the simulation. A correction factor must be applied to heat transfer calculations to avoid underestimating conduction.

(2) Convection

Convection in CFD-DEM refers to the heat transfer between fluid (gas or liquid) and solid particles due to their relative motion and temperature difference. It is a dominant mechanism in processes like fluidized bed reactors.

The convective heat transfer rate q_{if} between a particle i and the surrounding fluid is calculated as:

$$q_{if} = h_{\text{conv}} A_i (T_f - T_i) \tag{7.11}$$

where h_{conv} is the convective heat transfer coefficient; A_i is the surface area of particle i; T_f is the fluid temperature in the cell containing the particle; T_i is the temperature of particle i.

The convective heat transfer coefficient h_{conv} is influenced by several factors, such as the fluid's velocity, viscosity, thermal conductivity and whether the flow is laminar or turbulent. Typically, higher fluid velocities and turbulent flow increase the value of h_{conv}, leading to more efficient heat transfer. The coefficient h_{conv} is derived from the Nusselt number Nu correlation:

$$h_{\text{conv}} = \frac{Nu \cdot k_f}{d_p} \tag{7.12}$$

where k_f is the thermal conductivity of fluid; d_p is the particle diameter.

The Nusselt number Nu is often determined using empirical correlations, which depend on the flow regime (laminar or turbulent) and the Prandtl number Pr, a dimensionless number representing the ratio of momentum diffusivity to thermal diffusivity. The most utilized model is the Ranz-Marshall correlation [9]:

$$Nu = 2 + aRe^{\frac{1}{2}} Pr^{\frac{1}{3}} \tag{7.13}$$

where a is the empirical coefficient which is typically 0.6–1.2; Re is the particle Reynolds number; Pr is the local Prandtl number.

(3) Radiation

Radiation is the transfer of heat through electromagnetic waves, primarily in the form of infrared radiation, without the need for direct contact or a surrounding medium. In DEM, radiative heat transfer becomes significant when dealing with high-temperature processes, such as in combustion, sintering or metallurgical applications. Although it generally contributes less to heat transfer compared to conduction and convection, radiation can dominate in systems where the temperature differences are large or where the particles have high emissivity. The radiative heat transfer in DEM is influenced by several factors. The amount of radiative heat transfer increases significantly with temperature, following the Stefan-Boltzmann law, which is proportional to the fourth power of the absolute temperature. Larger particles with more surface area emit more radiation. The shape of particles can also affect the view factors, which determine the angle at which radiation is emitted and received. Additionally, the spatial arrangement of particles affects the line of sight between particles and thus the radiative heat transfer.

The continuum approach to modeling radiation in DEM treats the particulate system as a continuous medium, simplifying the complex interactions between individual particles. This approach is computationally less intensive than the discrete approach but may require assumptions and simplifications that can affect accuracy. In the continuum approach, the radiative heat transfer for a particle i immersed in this continuum radiation field can be calculated using a simplified form of the Stefan-Boltzmann law [10]:

$$q_{i,\mathrm{rad}} = \sigma \in A_p\left(T_R^4 - T_i^4\right) \tag{7.14}$$

where $\in$ is the emissivity of the particle surface, which depends on the material and surface characteristics; σ is the Stefan-Boltzmann constant of 5.67×10^{-8} W/m^{2}K^4; A_p is the surface area of the particle; T_i is the surface temperature of particle i. T_R is the local radiation temperature of the surrounding medium, which is determined by solving the radiative transfer equation for the entire domain.

The discrete approach, on the other hand, considers the radiative heat transfer between individual particles or between particles and reactor walls. This approach is more computationally demanding but provides a more accurate representation of the radiative heat transfer in particulate systems. In the discrete approach, the view factor F_{ij} plays a crucial role. It represents the fraction of radiation emitted by one particle that is directly received by another particle. The net radiative heat transfer from particle j to particle i can be calculated using:

$$q_{i,\mathrm{rad}} = \sigma \in_i \in_j A_i F_{ij}\left(T_j^4 - T_i^4\right) \tag{7.15}$$

where T_i and T_j are the temperatures of two interacting particles; F_{ij} is the view factor between particles i and j; A_i is the surface area of particle i, which is the receiving particle in the equation. Calculating view factors requires the information of the geometry and relative positions of the particles. Various numerical methods, such as the Monte Carlo method, can be used to determine view factors accurately. The total net radiative heat flow for particle i is then the sum of its exchanges with all other particles and walls in its line of sight.

(4) Heat Generation

In coupled CFD-DEM simulations, heat generation occurs through multiple physical and chemical processes that must be carefully modeled to achieve accurate thermal predictions. The primary mechanisms include chemical reaction heat, phase change heat, electrical/joule heating, etc. The chemical reaction heat is mainly considered in most CFD-DEM applications. Assuming that the rate is defined on the basis of particle surface, the generated heat for a reacting particle is calculated as:

$$q_{rxn} = A_i(-R_{rxn})(-\Delta H_{rxn}) \tag{7.16}$$

where A_i is the particle surface area; R_{rxn} is the reaction rate per unit area; ΔH_{rxn} is the heat of reaction. For an exothermic reaction, this term yields a positive heat source for the particle. For an endothermic reaction, it yields a negative source.

For gas–solid reactions like combustion and gasification, the reaction rate can be expressed by a general kinetic model:

$$R_{rxn} = k_{\text{kin}} \cdot f(C_i) \cdot g(T) \cdot h(X) \tag{7.17}$$

where k_{kin} is the intrinsic kinetic constant; $f(C_i)$ represents the dependence on the concentration of gaseous reactants; $g(T)$ represents Arrhenius-type temperature dependence; $h(X)$ is a function describing how the rate changes with the solid conversion degree X.

7.1.3 Coupling Strategies for Heat Transfer

The DEM-to-CFD transfer involves mapping particle-level heat fluxes to the Eulerian fluid grid, where each particle's contribution is carefully distributed to account for spatial resolution differences. For each CFD cell, particle heat fluxes are mapped to the Eulerian grid using [11]:

$$Q_{fp} = -\frac{1}{V_{\text{cell}}} \sum_{i=1}^{n_j} x_{A,i} q_{if} \tag{7.18}$$

where $x_{A,i}$ is the fraction of particle surface inside the cell, accounting for partial cell occupancy when particles straddle multiple cells. The reverse CFD-to-DEM coupling updates particle temperatures based on the surrounding fluid conditions using Eq. (7.1).

Temporal synchronization forms another critical aspect of the coupling framework. The inherent difference in timescales between the fast-moving particles (requiring small DEM time steps) and the relatively slower fluid dynamics using larger CFD steps necessitates specialized integration strategies. Common approaches include sub-cycling schemes with multiple DEM iterations per CFD step, or more sophisticated predictor–corrector methods that iteratively refine the coupled solution.

7.2 Fundamentals of Mass Transfer

Mass transfer in CFD-DEM involves tracking species transport between fluid and solid phases. The governing equations involved are divided into solid-phase and fluid-phase formulations.

7.2.1 Mass Transfer Mechanisms

Mass transfer in coupled CFD-DEM simulations refers to the physical processes that lead to the exchange of chemical species between the discrete solid particles and the surrounding continuous fluid phase. This exchange is a critical component in many industrial applications, such as combustion, gasification, drying, catalysis and mineral processing. The primary mechanisms governing this transfer are diffusion and convection, which are often driven by the chemical reactions occurring at the fluid–solid interface.

(1) Diffusion and Convection in Fluid Phase

Diffusion is the process by which species move from regions of high concentration to regions of low concentration due to random molecular motion. In the context of fluid-particle systems, diffusion governs the transport of reactants toward the particle surface and the transport of products away from it. The rate of diffusive mass transfer is quantified by Fick's Law and is highly dependent on the concentration gradient near the particle surface, as well as the effective diffusivity of the species in the fluid mixture.

Convection involves the bulk movement of fluid, which carries species toward or away from the particle surface. The convective flow can be driven by global pressure gradients (forced convection) or by local density variations due to temperature or composition differences (natural convection). The convective mass transfer rate is strongly influenced by the local fluid flow field around the particle, which is itself affected by the particle's geometry, velocity and the presence of other particles.

(2) Heterogeneous Reaction

In systems involving heterogeneous chemical reactions (e.g., char combustion, catalytic conversion), the reaction itself acts as a powerful driver for mass transfer. The consumption of a reactant at the particle surface creates a steep concentration gradient, which pulls more reactant toward the surface. Simultaneously, the generation of products at the surface creates a gradient that pushes them away. The reaction kinetics thus directly couples with and influences the diffusive and convective transport processes.

7.2.2 Governing Equations of Mass Transfer

(1) Solid-Phase Mass Balance

For processes where particles neither gain/lose mass nor change size, such as homogeneous reactions in the fluid phase or catalytic reactions where solids act only as stationary catalysts, the governing equations for particle dynamics retain their original form as specified in Chap. 2, without requiring additional mass-coupled terms. However, when the mass or size of particles changes due to reactions, the following modified Newton's second law is used [11]:

$$m_i \frac{\mathrm{d}\boldsymbol{v}_i}{\mathrm{d}t} = -\boldsymbol{v}_i R_{p,i} + \sum \boldsymbol{F}_i \tag{7.19}$$

where $R_{p,i}$ represents the rate of mass change of particle i; $\sum \boldsymbol{F}_i$ is the sum of all forces acting on the particle.

$$R_{p,i} = \frac{\mathrm{d}m_i}{\mathrm{d}t} \tag{7.20}$$

where $R_{p,i} > 0$ means mass loss and $R_{p,i} < 0$ means mass gain.

For angular momentum, the modified equation is expressed as [11]:

$$I_i \frac{\mathrm{d}\boldsymbol{\omega}_i}{\mathrm{d}t} = -\boldsymbol{\omega}_i \frac{\mathrm{d}I_i}{\mathrm{d}t} + \sum \boldsymbol{T}_i \tag{7.21}$$

where $\boldsymbol{\omega}_i$ is the angular velocity of particle i; I_i is the moment of inertia, and $\sum \boldsymbol{T}_i$ is the sum of all torques on particle i.

(2) Fluid-Phase Species Transport

The fluid phase tracks species concentration using a scalar transport equation [12]:

$$\frac{\partial\left(\alpha_f \rho_f Y_{f,i}\right)}{\partial t} + \nabla \cdot \left(\alpha_f \rho_f \boldsymbol{v}_f Y_{f,i}\right) = \nabla \cdot \left(\rho_f D_i \alpha_f \nabla Y_{f,i}\right) + S_{f,i} \tag{7.22}$$

where $Y_{f,i}$ is the mass fraction of species i in the fluid; D_i is the effective diffusivity including both viscous and turbulent contirbutions; $S_{f,i}$ is the source term which represents the mass production/consumption rate per unit volume for species i.

When the fluid phase experiences no net mass gain or loss, which means the total mass of all species remains constant, the standard CFD-DEM formulation needs to be supplemented solely with species conservation equations to account for compositional changes. However, when the mass is generated/consumed in the fluid, the continuity and momentum equations should be modified:

$$\frac{\partial\left(\alpha_f \rho_f\right)}{\partial t} + \nabla \cdot \left(\alpha_f \rho_f \boldsymbol{v}_f\right) = S_f \tag{7.23}$$

$$\frac{\partial\left(\alpha_f \rho_f \boldsymbol{v}_f\right)}{\partial t} + \nabla \cdot \left(\alpha_f \rho_f \boldsymbol{v}_f \otimes \boldsymbol{v}_f\right) = -\alpha_f \nabla p + \nabla \cdot \alpha_f \boldsymbol{\tau}_f + \alpha_f \rho_f \boldsymbol{g} + \boldsymbol{F}_{pf} + \boldsymbol{S}_F \tag{7.24}$$

where S_f and $\boldsymbol{S}_F$ account for the mass and momentum sources due to reactions respectively.

7.2.3 Coupling Strategies for Mass Transfer

The mass source term for fluid cells represents the net mass transfer from particles to the fluid in a cell. According to conservation of mass, it is expressed as:

$$S_f = -\frac{1}{V_{\text{cell}}} \sum_{i=1}^{n_j} R_{p,i} \tag{7.25}$$

The momentum source term accounts for the momentum transferred along with the mass exchange. The mass leaving or entering a particle carries the momentum of that particle at the moment of transfer. Therefore, it is calculated as:

$$\boldsymbol{S}_F = -\frac{1}{V_{\text{cell}}} \sum_{i=1}^{n_j} R_{p,i} \boldsymbol{v}_{p,i} \tag{7.26}$$

where $\boldsymbol{v}_{p,i}$ is the velocity vector of particle i. This term represents the rate of momentum injection into the fluid due to the mass efflux from the particles.

References

1. S. Wang and Y. Shen, CFD-DEM modelling of dense gas-solid reacting flow: Recent advances and challenges, Prog Energy Combust Sci. **109**, 101–221 (2025). https://doi.org/10.1016/j.pecs.2025.101221
2. Y. He, A. Ren, T. Tang, T. Wang, Multi-scale numerical simulation of flow, heat and mass transfer behaviors in dense gas-solid flows: A brief review, May 01, (2022). Science Press. https://doi.org/10.1007/s11630-022-1605-x.
3. T. Brosh and A. Levy, Modeling of heat transfer in pneumatic conveyer using a combined DEM-CFD numerical code, Dry Technol, **28**(2), 155–164 (2010). https://doi.org/10.1080/07373930903517482
4. S. Yagi, D. Kunii, Studies on effective thermal conductivities in packed beds, AIChE J, **3**(3), 373–381 (1957). https://doi.org/10.1002/aic.690030317
5. Z.Y. Zhou, A.B. Yu, P. Zulli, Particle scale study of heat transfer in packed and bubbling fluidized beds, AIChE J, **55**(4), 868–884 (2009). https://doi.org/10.1002/aic.11823
6. G.K. Batchelor, R. O'brien, Thermal or electrical conduction through a granular material, Proc Roy Soc Lond A Math Phys Sci, **355**, 313–333 (1977)
7. J.H. Zhou, A.B. Yu, M. Horio, Finite element modeling of the transient heat conduction between colliding particles, Chem Eng J, **139**(3), 510–516 (2008). https://doi.org/10.1016/j.cej.2007.08.024
8. Z.Y. Zhou, A.B. Yu, P. Zulli, A new computational method for studying heat transfer in fluid bed reactors, Powder Technol, **197**(1), 102–110 (2010). https://doi.org/10.1016/j.powtec.2009.09.002
9. W. Ranz, W. Marshall, Evaporation from droplets, Chem Eng Prog, **48**, 141–146 (1952)
10. H. Zhou, G. Flamant, D. Gauthier, Y. Flitris, Simulation of coal combustion in a bubbling fluidized bed by distinct element method, Chem Eng Res Des, **81**(9), 1144–1149 (2003). https://doi.org/10.1205/026387603770866308
11. H.R. Norouzi, R. Zarghami, R. Sotudeh-Gharebagh, N. Mostoufi, Coupled CFD-DEM modeling: formulation, implementation and application to multiphase flows. John Wiley & Sons, (2016)
12. D.M. Snider, S.M. Clark, P.J. O'Rourke, Eulerian-Lagrangian method for three-dimensional thermal reacting flow with application to coal gasifiers, Chem Eng Sci. **66**(6), 1285–1295 (2011). https://doi.org/7.110.1016/j.ces.2010.12.042

Chapter 8
Strategies for Large-Scale CFD-DEM Simulations

8.1 Supercomputing Hardware

Supercomputing hardware forms the backbone of large-scale simulations. The simulations, involving complex multiphase flows, particle dynamics and interactions with fluid media, require significant computational power [1]. The hardware employed must be capable of handling massive datasets and performing numerous calculations simultaneously. The choice of hardware directly impacts the speed, accuracy and feasibility of such simulations. The most commonly used supercomputing hardware for CFD-DEM simulations includes central processing units (CPUs), graphics processing units (GPUs) and specialized systems such as clusters and supercomputers [2]. Each hardware platform offers distinct advantages and can be selected based on the scale of the problem being solved, the computational efficiency required and the available budget.

8.1.1 Central Processing Units (CPUs)

CPUs, the traditional workhorses of computing, are designed for serial processing, meaning they perform tasks sequentially. While modern CPUs contain multiple cores, they are still more limited in their ability to handle parallel tasks compared to GPUs. CPUs are ideal for tasks that require high single-threaded performance, such as preprocessing data or handling complex logic. However, for CFD-DEM simulations, where thousands or even millions of particles interact with the surrounding fluid, CPUs alone are often not fast enough. The inherent parallelism of particle-pair interactions and fluid-particle coupling makes it a natural fit for more parallel computing architectures like GPUs. Despite this, CPUs are still commonly used in smaller simulations or in scenarios where the full parallelization potential of a GPU cannot be fully realized.

R. K. Agarwal and Y. Shao, *Theory and Application of Discrete Element Method to Simulation of Multiphase Flows*, https://doi.org/10.1007/978-3-032-11157-9_8

8.1.2 Graphics Processing Units (GPUs)

GPUs have become a cornerstone for accelerating large-scale simulations. Originally designed for rendering complex images in video games, GPUs are now extensively employed for parallel computation due to their highly parallel architecture, making them ideal for handling the vast computational demands of CFD-DEM simulations. GPUs consist of thousands of smaller cores that can handle multiple tasks simultaneously, which contrasts with the architecture of CPUs, where cores are designed for high-speed sequential processing. In CFD-DEM simulations, this parallel processing capability allows for the simultaneous evaluation of forces and particle interactions in a system. For example, when solving the governing equations for fluid-particle interactions, each particle's position, velocity and force calculations can be processed in parallel. This drastically reduces the computation time, enabling simulations of much larger systems and more complex interactions. Furthermore, modern GPUs come with specialized libraries and frameworks, such as NVIDIA's CUDA (Compute Unified Device Architecture), which enables developers to write code that takes full advantage of GPU capabilities. CUDA allows for the parallelization of tasks across many cores, which is essential for speeding up the numerical methods used in CFD-DEM. Additionally, GPUs offer higher memory bandwidth, which enables faster data access and transfer, a critical factor when dealing with the large datasets generated in CFD-DEM simulations.

In practice, using GPUs for CFD-DEM simulations requires careful optimization. The computational performance can be limited by several factors, such as data transfer between the CPU and GPU, memory constraints and the efficiency of the parallel algorithms used. For instance, in a CFD-DEM simulation, if the particle interactions are not efficiently distributed across the GPU cores, or if there is significant data movement between the CPU and GPU, the speed-up from GPU usage may not be as significant as expected. To mitigate these challenges, hybrid computing approaches that combine CPUs and GPUs are often used, where the GPU handles the more computationally intensive tasks such as particle dynamics, while the CPU manages other tasks, such as preprocessing and post-processing.

8.1.3 High-Performance Clusters and Supercomputers

In addition to individual CPUs and GPUs, many large-scale simulations are run on high-performance computing (HPC) clusters or supercomputers, which are collections of interconnected processing nodes. These systems can contain a mix of CPUs, GPUs and specialized hardware designed to work together to solve complex problems. For example, the Oak Ridge National Laboratory's Summit supercomputer, one of the world's most powerful supercomputers, combines IBM POWER9 CPUs and NVIDIA V100 GPUs to deliver immense computational power. In HPC clusters, simulations are typically broken into smaller tasks and distributed across many nodes, each equipped with its own processor. These nodes communicate over high-speed networks to exchange information and synchronize results. HPC clusters enable extremely large simulations, such as those used in climate modeling, material science and large-scale fluid dynamics. The major challenge when using HPC clusters is efficiently distributing the workload. This can be done through parallelization strategies, which ensure that the computational resources of the entire cluster are used effectively. These strategies are covered in detail in later sections, which include techniques such as particle-level parallelization, time-step parallelization and spatial domain decomposition.

8.1.4 Cloud Computing and Distributed Systems

In addition to traditional supercomputing facilities, cloud computing has emerged as a flexible, cost-effective solution for large-scale simulations. This allows researchers to scale their computational resources based on the requirements of their simulation, avoiding the need for a large, upfront investment in hardware. Cloud computing also enables the use of distributed systems, where tasks are spread across multiple machines or data centers. This is particularly useful for applications such as CFD-DEM, where simulations can be easily parallelized across multiple nodes or GPUs. While cloud computing offers scalability and flexibility, it does come with challenges such as network latency, data security and potential data transfer costs, which need to be carefully managed for efficient simulation performance.

In summary, the hardware for supercomputing plays a critical role in the success of large-scale CFD-DEM simulations. From CPUs and GPUs to high-performance clusters and cloud computing, the choice of hardware affects the speed, scalability and feasibility of these simulations. With advancements in GPU technology and the rise of cloud-based supercomputing platforms, researchers are now able to perform simulations that were previously unthinkable, opening up new possibilities for understanding complex fluid-particle interactions in a wide range of industrial applications.

8.2 Parallelization Strategies

In large-scale CFD-DEM simulations, parallelization plays a crucial role in reducing computation time and enabling the simulation of complex systems with a large number of particles and fluid elements. There are various parallelization strategies that can be employed to efficiently distribute the workload across multiple processors, CPUs, GPUs or computational nodes. The following sections introduce four key parallelization strategies commonly used in CFD-DEM simulations: particle-level parallelization, time-step parallelization, spatial domain decomposition parallelization and GPU parallelization.

8.2.1 Temporal Domain Parallelization

Temporal domain parallelization is another powerful strategy for accelerating large-scale CFD-DEM simulations. This method focuses on parallelizing the process of advancing the system through multiple time steps. In a typical CFD-DEM simulation, the system's state is updated at discrete time intervals. These updates involve complex calculations for particle interactions, fluid dynamics and the coupling between the two. At each time step, the state of the system is updated based on the governing equations. The time-step parallelization strategy divides the task of updating the system state into smaller chunks of time steps, allowing different computational units to handle different parts of the temporal progression. Rather than processing all the particles and fluid interactions sequentially, multiple processors can calculate the evolution of the system over different time steps in parallel.

In modern high-performance computing environments, hybrid CPU-GPU systems are commonly used to exploit both the computational power of CPUs and the parallel processing capabilities of GPUs. Time-step parallelization can take full advantage of such hybrid systems by dividing the work within each time step across different types of processors. For instance, in a simulation where the gas phase is solved using a CFD solver on the CPU and the particle dynamics are handled by the DEM solver on the GPU, time-step parallelization can allow the CPU and GPU to operate concurrently during each time step. While the CPU calculates the fluid dynamics for a given time step, the GPU can simultaneously compute the particle dynamics, and the two results are then synchronized to update the system state for the next time step. In this way, time-step parallelization across CPU and GPU units not only accelerates the simulation but also optimizes the use of each processing unit's strengths.

While time-step parallelization can yield substantial performance improvements, there are several challenges that need to be addressed. One of the primary challenges is ensuring correct synchronization between processors. Since the time-step progression is sequential, with each step depending on the results of the previous one, the processors must coordinate and share intermediate data to maintain consistency across the simulation. This can introduce communication overhead, especially if the time-step intervals are large or if there are frequent interactions between particles and the fluid. Moreover, for systems that experience rapid changes in behavior, the optimal time step might vary throughout the simulation. This dynamic adjustment of time steps can complicate parallelization, as it requires careful handling to ensure that different processors are working with consistent data and that the simulation can adapt to different timescales across different regions of the system. Another challenge lies in the computational resources required for time-step parallelization. For large systems, dividing the time-step updates across many processors requires significant memory bandwidth, and efficient memory management is crucial. If the memory is not optimally distributed, the simulation might encounter bottlenecks that reduce the potential speed-up from parallelization.

8.2.2 Spatial Domain Decomposition Parallelization

The key idea behind spatial domain decomposition is to break the simulation domain into smaller, non-overlapping regions or subdomains. These subdomains are then assigned to different processors or computing units, which are responsible for solving the equations governing the system's behavior within their assigned region. The subdomains may be in the form of cubes, rectangular prisms or other shapes, depending on the geometry of the physical system being simulated.

One important consideration in spatial domain decomposition is the handling of interactions between particles and between particles and the fluid across the boundaries of adjacent subdomains. Since the behavior of particles often depends on their interactions with nearby particles, information must be exchanged between neighboring subdomains. This is typically done through communication between processors, where each processor sends and receives data related to particle positions, velocities, forces or other relevant properties from its neighboring subdomains. This communication step is crucial for ensuring the accuracy and consistency of the simulation, especially when particles or fluid flow cross the boundaries of subdomains.

8.3 Numerical Acceleration Technique

8.3.1 Coarse-Grained Method

The coarse-grained method (CGM) is an important numerical acceleration technique designed to overcome the primary computational bottleneck in CFD-DEM simulations: the exorbitant cost of tracking every individual particle in systems containing millions or even billions of granules [3]. The fundamental principle of CGM is to reduce the number of computational elements by grouping multiple real particles that share similar properties such as velocity, temperature and chemical composition into a single, representative numerical entity often called a "parcel" or "coarse-grained particle," as shown in Fig. 8.1. This parcel then moves and interacts with the fluid and other parcels as a surrogate for the entire group of real particles it represents. The core idea is that while the microscopic details of each particle are lost, the macroscopic behavior of the system, which is often the primary interest in engineering applications, is preserved.

The mathematical foundation of the CGM rests on the concept of a statistical weight, denoted as W. This scalar value W defines the number of real particles contained within one coarse-grained parcel [4]. All physical properties and interaction forces associated with the parcel are scaled by this weight to ensure conservation of mass, momentum and energy at the macroscopic scale. For example, the mass of a parcel is W times the mass of a single real particle. The interphase coupling terms

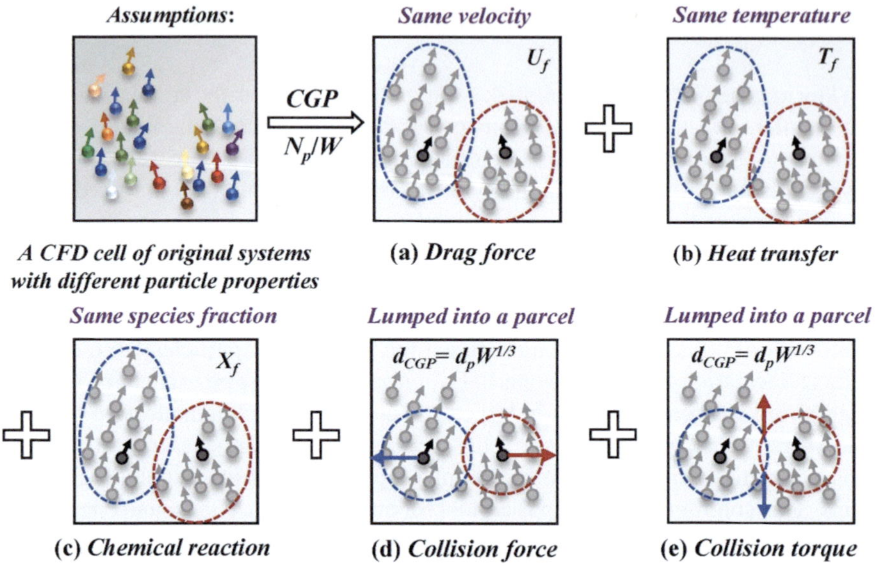

Fig. 8.1 Basic assumptions in the coarse-grained method (CGM) [1]

are calculated by summing the scaled contributions from all parcels within a computational cell. The drag force exerted on the fluid phase, for instance, is formulated as the sum of the drag on each parcel, which itself represents the cumulative drag of W particles. Similarly, heat transfer rates and species reaction rates are computed using this weighted-sum approach.

The implementation of CGM yields dramatic improvements in computational efficiency. However, the method introduces a trade-off between computational speed and resolution. The most significant limitation is the loss of fine-grained, particle-scale information. The choice of the coarse-grained ratio W is therefore a critical compromise. It is generally recommended that W be kept below a value of 5 to accurately capture key physics such as particle–wall interactions, cohesive forces, and detailed mixing and segregation patterns. Furthermore, a high W value means that the parcel represents a larger physical volume. This can conflict with a standard requirement in unresolved CFD-DEM that the fluid grid cell should be larger than the particle (or parcel) diameter to avoid numerical instabilities during the voidage calculation. To circumvent this limitation, the CGM is often coupled with advanced mapping algorithms, such as the statistical kernel or diffusion-based method. This powerful combination decouples the parcel size from the fluid grid size, allowing the fluid mesh to be refined to resolve flow structures without being constrained by the size of the coarse-grained parcels. This synergy enhances both the versatility and the accuracy of the CGM, solidifying its role as an indispensable tool for simulating complex, large-scale dense gas–solid reacting flows.

8.3.2 Recurrence CFD Method

The recurrence CFD (rCFD) method is a data-driven framework designed to overcome the high computational cost of simulating long-term physical processes in complex, unsteady flow systems. Unlike traditional CFD, which computes the flow field at every time step, rCFD operates on a fundamentally different principle. It recognizes that many industrially relevant flows, while turbulent and seemingly chaotic, often exhibit pseudo-periodicity, a tendency to revisit similar, albeit not identical, dynamic states over time [5]. This means the flow is characterized by recurring patterns, such as the cyclic formation, rise and eruption of bubbles in a fluidized bed. rCFD capitalizes on this inherent recurrence by extracting these patterns from a short, high-fidelity "training" simulation and then using them to reconstruct the flow's long-term evolution with minimal computational effort [6].

The method is based on the idea of recurrence analysis. It starts with a short, high-resolution simulation of the flow until it reaches a pseudo-steady state. During this simulation, snapshots of the entire flow field called recurrence frames are saved at regular intervals. A special metric, called a recurrence norm, is used to compare all these saved frames and measure how similar they are to each other. The results are organized into a recurrence matrix, which acts like a map showing which flow states look alike. This makes it possible to find places where different sequences of flow can be connected smoothly, extending the simulation far into the future without additional costly computation. This idea is used in two ways: flow-based rCFD and transport-based rCFD [7]. The flow-based rCFD uses the recurrence matrix to stitch together flow sequences from the saved database into one long, continuous artificial flow field. Then, passive elements like heat or chemicals are simulated as they move through this pre-built flow. The transport-based rCFD is more efficient which skips rebuilding the whole flow field. Instead, it records how imaginary tracer particles move between grid cells during the original short simulation. To predict long-term behavior, it simply reapplies these recorded "cell shifts" to the substance being tracked. This avoids solving complex equations repeatedly and is extremely fast.

The advantages of rCFD are profound, primarily its computational speed-up, which makes previously inaccessible long-time simulations feasible. However, its application is bound by critical limitations. Its core reliance on pseudo-periodicity means it is unsuitable for transient start-up, shut-down or highly irregular processes. Furthermore, it operates under a passive scalar assumption, which means the transported quantities must not significantly alter the underlying flow dynamics from which the recurrence patterns are derived. If a transported quantity exerts strong feedback on the flow, the recurrence statistics become invalid, and the method requires a new base simulation. Despite these constraints, recurrence CFD represents a paradigm shift from a purely equation-solving approach to a data-driven, pattern-recognition strategy, offering a powerful key to unlocking the simulation of extended timescales in complex industrial reactors.

References

1. L. Lu, S. Benyahia, Chapter two - advances in coarse discrete particle methods with industrial applications, in *Advances in Chemical Engineering*, vol. 53, ed. by A. Parente and J. De Wilde (Academic Press, 2018), pp. 53–151. https://doi.org/10.1016/bs.ache.2017.12.001
2. H.R. Norouzi, R. Zarghami, N. Mostoufi, New hybrid CPU-GPU solver for CFD-DEM simulation of fluidized beds. Powder Technol, **316**, 233–244 (2017). https://doi.org/10.1016/j.powtec.2016.11.061
3. S. Wang, Y. Shen, Coarse-grained CFD-DEM modelling of dense gas-solid reacting flow. Int J Heat Mass Transf, **184**, 122302 (2022). https://doi.org/10.1016/j.ijheatmasstransfer.2021.122302
4. S. Wang, Y. Shen, CFD-DEM modelling of dense gas-solid reacting flow: Recent advances and challenges. Prog Energy Combust Sci, **109**, 101221 (2025). https://doi.org/109101221-10.1016/j.pecs.2025.101221
5. P. Kieckhefen, T. Lichtenegger, S. Pietsch, S. Pirker, S. Heinrich, Simulation of spray coating in a spouted bed using recurrence. CFD Particuology, **42**, 92–103 (2019). https://doi.org/10.1016/j.partic.2018.01.008
6. T. Lichtenegger, E.A.J.F. Peters, J.A.M. Kuipers, S. Pirker, A recurrence CFD study of heat transfer in a fluidized bed. Chem Eng Des, **172**, 310–322 (2017). https://doi.org/10.1016/j.ces.2017.06.022
7. S. Pirker, T. Lichtenegger, Efficient time-extrapolation of single- and multiphase simulations by transport based recurrence CFD (rCFD). Chem Eng Des, **188**, 65–83 (2018). https://doi.org/10.1016/j.ces.2018.04.059

Chapter 9
Applications to Conventional Fluidized Beds

Conventional fluidized beds, including bubbling, spouted and circulating fluidized beds, are widely used in industries such as chemical processing, energy production, pharmaceuticals and biomass conversion due to their efficient particle mixing, excellent heat and mass transfer and uniform temperature distribution. The CFD-DEM approach is a powerful tool for studying fluidized beds, as it provides detailed insights into particle-scale dynamics, including mixing, segregation, heat transfer and chemical reactions. This approach helps optimize reactor design, improve process efficiency and reduce operational costs by accurately modeling the interplay between fluid dynamics and particle behavior, ultimately enhancing the performance of fluidized bed systems in industrial applications.

9.1 Bubbling Fluidized Bed

9.1.1 Research Focus

A bubbling fluidized bed (BFB) is a fluid-particle system in which a gas flows upward through a bed of solid particles at a velocity exceeding the minimum fluidization velocity but below the terminal velocity for particle entrainment. The excess gas forms voids or bubbles that rise through the bed, creating a two-phase system, including a dense/emulsion phase where particles remain closely packed and a bubble phase with low solids concentration. Figure 9.1 displays the CFD-DEM research framework for BFBs, which mainly focuses on bubble dynamics, particle–fluid interaction, heat and mass transfer, multiscale analysis, industrial application, etc.

Bubble dynamics are critical in BFBs since they directly determine the hydrodynamic behavior. Researchers systematically examine bubble nucleation, growth and

R. K. Agarwal and Y. Shao, *Theory and Application of Discrete Element Method to Simulation of Multiphase Flows*, https://doi.org/10.1007/978-3-032-11157-9_9

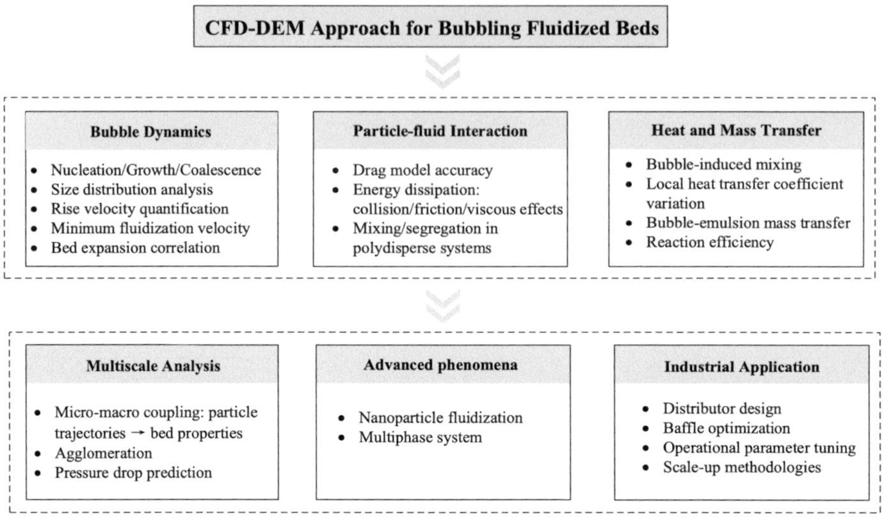

Fig. 9.1 CFD-DEM research framework for BFBs

coalescence patterns, with particular emphasis on quantifying bubble size distributions and rise velocities under various operating conditions. These bubble characteristics directly influence the transition between different fluidization regimes, particularly the critical minimum fluidization velocity that marks the onset of bubbling behavior. Furthermore, the relationship between bubble motion and bed expansion characteristics provides crucial insights into voidage distribution patterns and overall bed stability.

Building upon this hydrodynamic foundation, studies delve into the intricate particle–fluid interactions that drive system behavior. A key focus involves evaluating the accuracy of different drag models in capturing the momentum exchange between phases, which significantly affects bubble formation and particle motion. The analysis extends to energy dissipation mechanisms, where researchers quantify the relative contributions of particle collisions, frictional contacts and viscous dissipation to overall system dynamics. These investigations naturally lead to examinations of particle mixing and segregation phenomena, particularly in polydisperse systems where particle size and density variations can dramatically affect bed performance.

The understanding of these fundamental processes enables detailed investigations of heat and mass transfer, which are critical for industrial applications. Researchers analyze how bubble-induced mixing enhances convective heat transfer between phases, with particular attention to local heat transfer coefficients and their spatial variation. In chemical reaction systems, studies focus on the interplay between bubble dynamics and reaction efficiency, examining how bubble-mediated gas exchange affects conversion rates in catalytic and combustion processes. These transport phenomena are further complicated by mass transfer limitations between bubble and emulsion phases, which often represent the rate-limiting step in reactor performance.

To connect these observations across scales, multiscale analysis approaches have been developed. These frameworks systematically relate microscopic particle behavior, such as contact forces and trajectory patterns, to macroscopic bed properties like pressure drop and expansion ratios. Special attention is given to particle-level phenomena, including agglomeration tendencies and electrostatic effects, which can significantly impact bed stability and operability. Such comprehensive analysis provides the foundation for industrial applications, where research focuses on optimizing operational parameters and developing scale-up methodologies.

The practical implications of these studies are particularly evident in equipment design and process optimization. Investigations into distributor plate configurations and internal baffle arrangements demonstrate how strategic design modifications can improve bubble distribution and particle mixing. Process optimization studies systematically evaluate the effects of gas velocity, particle characteristics and bed geometry on overall system performance. These industrial applications benefit from specialized studies examining more complex scenarios, such as nanoparticle fluidization where interparticle forces dominate or multiphase systems incorporating liquid phases.

9.1.2 Example Applications

Zhu et al. [1] carried out a comprehensive CFD-DEM study on the hydrodynamics of Geldart B and D particles in a pressurized bubbling fluidized bed (PBFB). The research aims to elucidate the effects of operating pressure, particle size and initial bed height on key fluidization parameters, including minimum fluidization velocity U_{mf}, particle dynamics and bubble behavior. The simulations reveal that U_{mf} decreases with increasing pressure, particularly at lower pressures (0.1–0.4 MPa), and increases with larger particle sizes. Particle hydrodynamics are analyzed in detail, showing that particles exhibit upward motion in the bed's central region and downward backmixing near the walls, forming an internal circulation pattern. Elevated pressure leads to reduced bed expansion and more uniform radial particle distribution due to smaller and more frequent bubbles that restrict particle movement, as shown in Fig. 9.2. Smaller bubbles at higher pressures also result in lower particle velocities and a denser particle phase near the bed bottom. Bubble behavior is another critical focus, with the study highlighting the influence of pressure and particle size on bubble diameter, aspect ratio and density. Higher pressures promote bubble breakup, yielding smaller bubbles with higher aspect ratios and increased bubble density, as shown in Fig. 9.3. In contrast, larger particles facilitate the formation of fewer but larger bubbles or slugs, which enhance bed expansion. The initial bed height also plays a role, with taller beds favoring bubble coalescence and slug flow, particularly for larger particles.

Korkerd et al. [2] explored the use of BFBs for biomass combustion, focusing on enhancing mixing efficiency and heat transfer through the integration of immersed tubes, as shown in Fig. 9.4. Biomass, as a renewable energy source, presents challenges due to its low energy density and seasonal variability, prompting the use

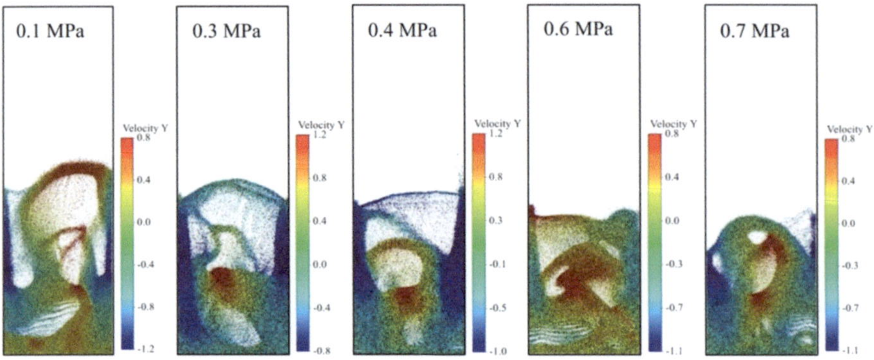

Fig. 9.2 Instantaneous particle distribution under different pressures [1]

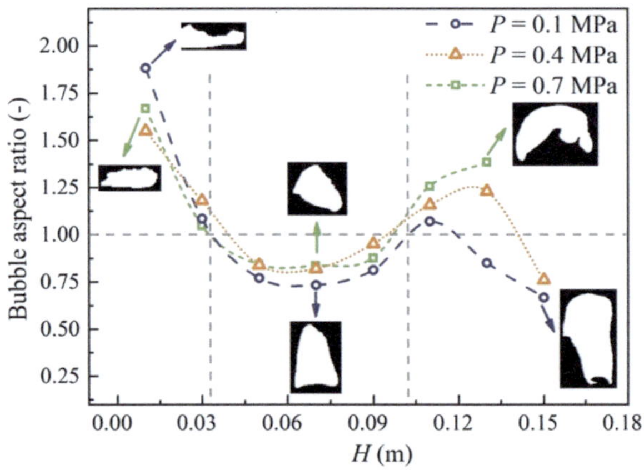

Fig. 9.3 Bubble aspect ratio versus height under different operating pressures [1]

of mixed biomass blends to ensure consistent fuel supply. However, the irregular shapes and sizes of biomass particles complicate fluidization dynamics, necessitating the addition of inert materials like silica sand to improve fluidization and heat transfer. Despite these adjustments, the inherent complexity of biomass-sand mixtures requires further investigation into their hydrodynamic and thermal behaviors. The CFD-DEM simulations are conducted to study particle dynamics and heat transfer in BFBs. Studies demonstrate that staggered tube arrangements significantly enhance mixing by promoting turbulence and disrupting bubble coalescence, leading to more uniform particle distribution compared to in-line configurations. Smaller tube diameters further improve mixing by increasing particle–particle collisions, while larger diameters enhance heat transfer due to greater surface area for thermal exchange. Statistical approaches, such as the 2^K factorial design and Analysis of Variance, are employed to systematically evaluate design parameters. Results indicate that tube diameter and the angle between tubes are the most influential factors, with interactions between these parameters also playing a critical role. Heat transfer

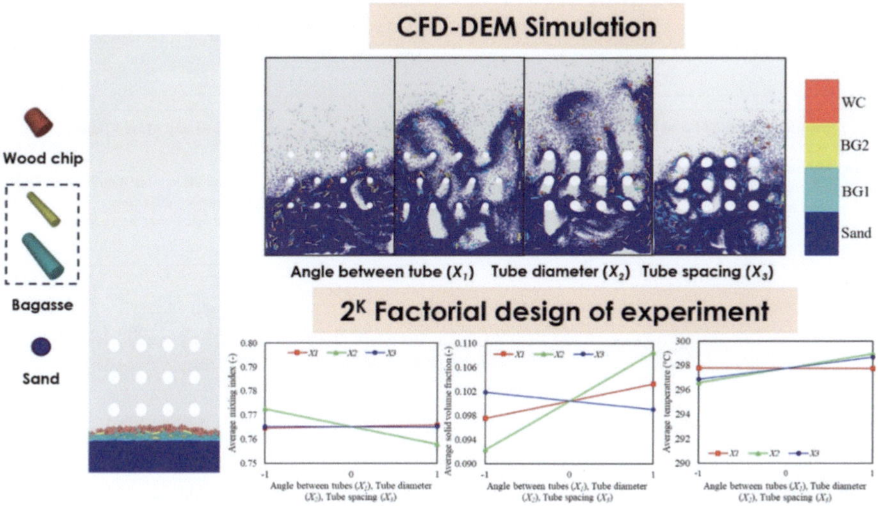

Fig. 9.4 Effect of immersed tubes configurations on mixing and heat transfer of mixed biomass and silica sand in a BFB [2]

studies reveal that conduction dominates in the thermal boundary layer, with convective heat transfer contributing significantly in regions of high particle concentration. The spatial distribution of heat transfer coefficients shows higher values around staggered tubes, attributed to increased particle-tube interactions and improved flow dynamics.

9.2 Spouted Fluidized Bed

9.2.1 Research Focus

A spouted fluidized bed (SFB) operates by injecting a high-velocity gas stream centrally through a nozzle at the bed's base, creating three distinct flow regimes: spout, annulus and foundation. The spout is formed by a high-velocity and gas-rich central jet carrying particles upward. The annulus is formed by a surrounding region where particles descend slowly in a packed or mildly fluidized state. The fountain is a particle spray at the bed's top, where spout-entrained particles disengage and fall back to the annulus. Unlike traditional BFBs, SFBs rely on cyclic particle circulation (spout → fountain → annulus → spout) to achieve mixing and heat/mass transfer.

Figure 9.5 displays the CFD-DEM research framework for SFBs. A primary research focus of SPBs involves the fundamental mechanisms of spout formation and stability. Studies systematically examine the minimum spouting velocity and its

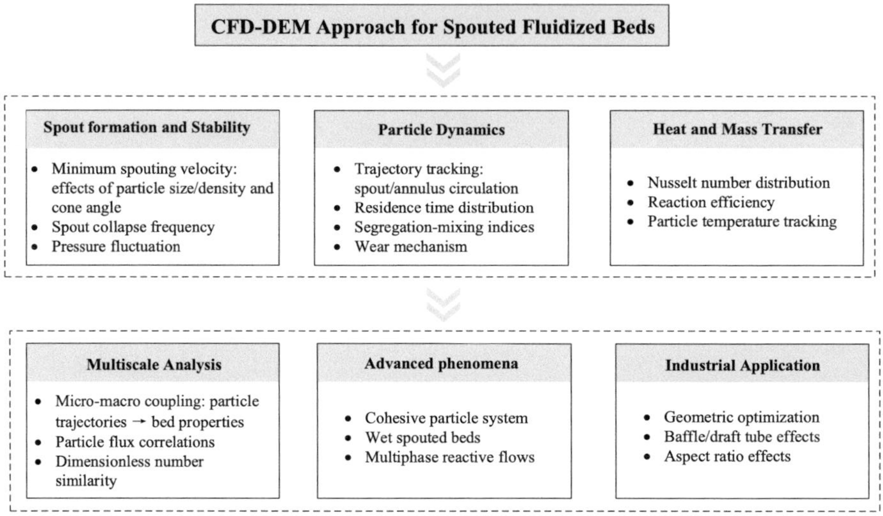

Fig. 9.5 CFD-DEM research framework for SFBs

dependence on particle properties such as size and density, as well as bed geometry parameters including cone angle and inlet diameter. Advanced experimental techniques like high-speed imaging and frequency domain analysis are employed to characterize spout dynamics, examining phenomena such as spout collapse and pressure fluctuations.

The particle motion and mixing characteristics represent another critical research area. Individual particle trajectories are tracked to distinguish between upward-moving particles in the spout zone and downward-flowing particles in the annular region. Quantitative analysis of residence time distributions offers valuable information about particle circulation patterns, while studies on multicomponent systems focus on segregation-mixing dynamics. By developing mixing indices correlated with operational parameters, researchers can optimize process efficiency. Additionally, particle wear mechanisms are investigated through contact force statistics, particularly in high-shear zones at the spout-annulus interface, which is essential for industrial applications involving abrasive materials. Additionally, the gas flow pattern evolution is carefully characterized, including jet penetration and gas exchange between spout and annulus.

Heat and mass transfer processes in SFBs receive considerable attention due to their industrial relevance. Researchers derive local Nusselt number distributions from coupled particle-gas temperature calculations to guide thermal process optimization. For reactive systems, investigations focus on how spout structure enhances reactant mixing and conversion efficiency, particularly in applications like biomass gasification. By tracking particle temperature histories during circulation, studies assess heat loss mechanisms and thermal uniformity, which are vital for scaling up processes.

Equipment design and scale-up studies form a critical application-oriented research direction. Parametric studies systematically evaluate the effects of geometric parameters such as cone angle, inlet diameter and bed aspect ratio on spouting stability and particle circulation rates. Furthermore, scale-up criteria are developed through multiscale simulations, establishing similarity principles based on dimensionless numbers or particle flux. These efforts bridge the gap between laboratory-scale experiments and industrial-scale operations, ensuring reliable and efficient scale-up of SFB technology.

9.2.2 Example Applications

Luo et al. [3] conducted a comprehensive CFD-DEM study on the spout deflection behavior of cohesive particles in a SPF, which is a critical instability issue affecting the operational range and efficiency of fluidized beds, particularly when handling cohesive particles like those covered with water or honey. Three types of particles including dry particles, water-covered cohesive particles and honey-covered cohesive particles are investigated, with the snapshots displayed in Fig. 9.6. Key findings show that cohesive particles exhibit smaller spout deflection angles compared to dry particles. The viscosity and initial liquid content of the cohesive particles significantly influence spout deflection, with higher liquid content reducing the median spout deflection angle. Additionally, the results reveal that while the vertical pressure drop remains consistent across particle types, horizontal pressure distribution is highly sensitive to liquid content, liquid type and spouting gas velocity.

Luo et al. [4] further conducted a follow-up study to the previous work on spout deflection in fluidized beds, now focusing on non-spherical particles such as cubic, ellipsoid, and cylindrical shapes using a superquadric CFD-DEM model. Figure 9.7 displays the instantaneous distribution of particles with different shapes. Results show that the spout channel narrows for non-spherical particles, with cubic particles exhibiting the narrowest channel and the most significant spout deflection. Cylindrical particles show the weakest deflection due to their interlocking behavior, while ellipsoid particles display intermediate behavior with occasional spout incoherence. The maximum lift height of particles varies with shape, with cubic particles being the easiest to fluidize. Additionally, particle orientation analysis reveals that cubic particles tend to form ordered arrangements near walls, while other shapes distributed more randomly.

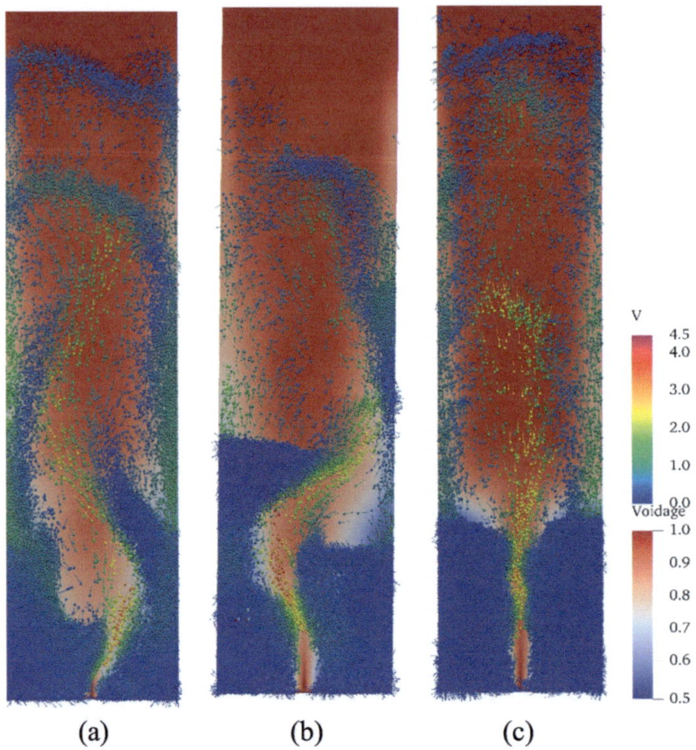

Fig. 9.6 Snapshot of **a** dry particles; **b** cohesive particles with water; **c** cohesive particles with honey [3]

9.3 Circulating Fluidized Bed

9.3.1 Research Focus

A circulating fluidized bed (CFB) is an advanced gas–solid reactor system characterized by high-velocity gas flow that entrains solid particles, creating a continuous circulation loop between the riser and a particle separator. Unlike bubbling or spouted beds, CFBs operate at gas velocities significantly exceeding the terminal velocity of particles, resulting in dilute-phase suspension flow in the riser, high particle flux with rapid recirculation and enhanced mixing and heat/mass transfer due to intense turbulence. Figure 9.8 displays the CFD-DEM research framework for CFB.

A primary research focus on CFBs lies in characterizing gas–solid flow regimes and regime transitions. Studies systematically examine the transition from bubbling to fast fluidization, determining critical velocities such as minimum fluidization velocity and transport velocity. The dynamic formation and disintegration of particle clusters in both dilute and dense phases are analyzed, with particular attention to

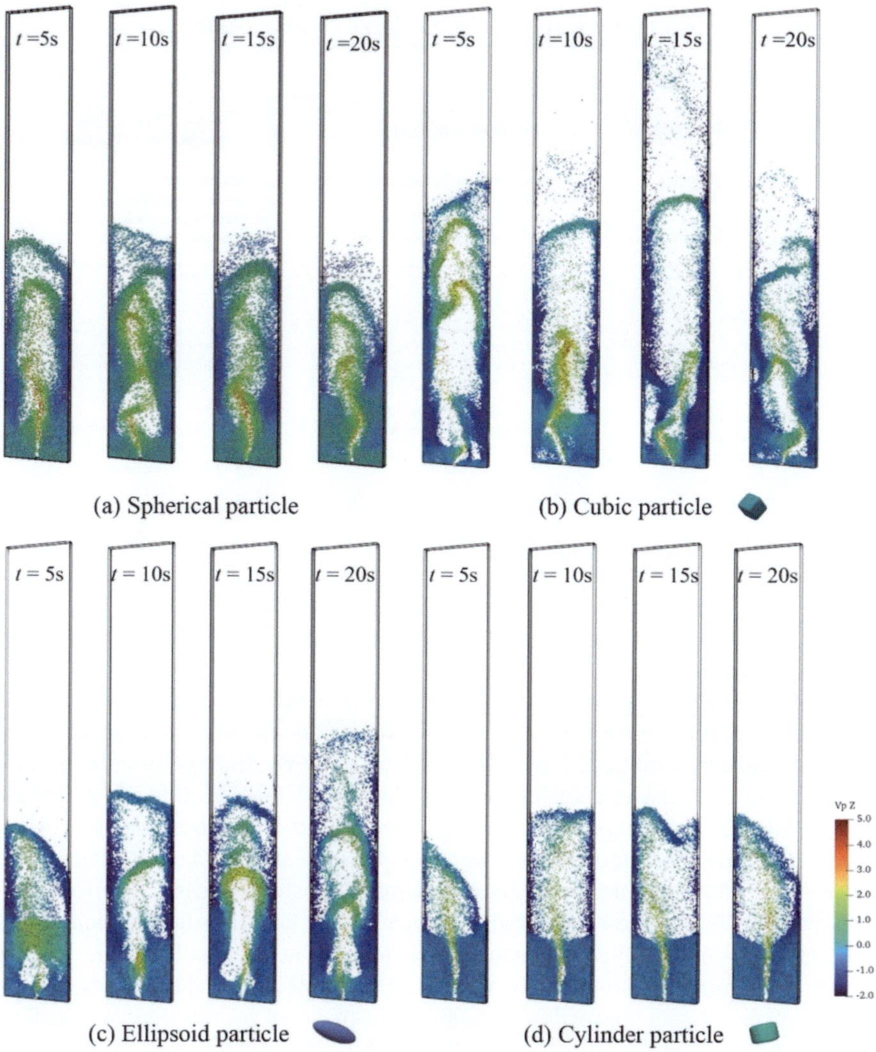

Fig. 9.7 Instantaneous distribution of particles colored by the particle vertical velocity: **a** spherical particles; **b** cubic particles; **c** ellipsoid particles; **d** cylinder particles [4]

their impact on flow stability. Axial pressure profiles are quantified to establish relationships between operating parameters and system hydrodynamics.

The particle motion and circulation patterns represent another key research area. CFD-DEM simulations track individual particle trajectories through the riser, separator and return system, providing statistics on particle residence times and circulation frequencies. Special emphasis is placed on return system stability, investigating flow obstruction risks in standpipes and loop seals through detailed force network

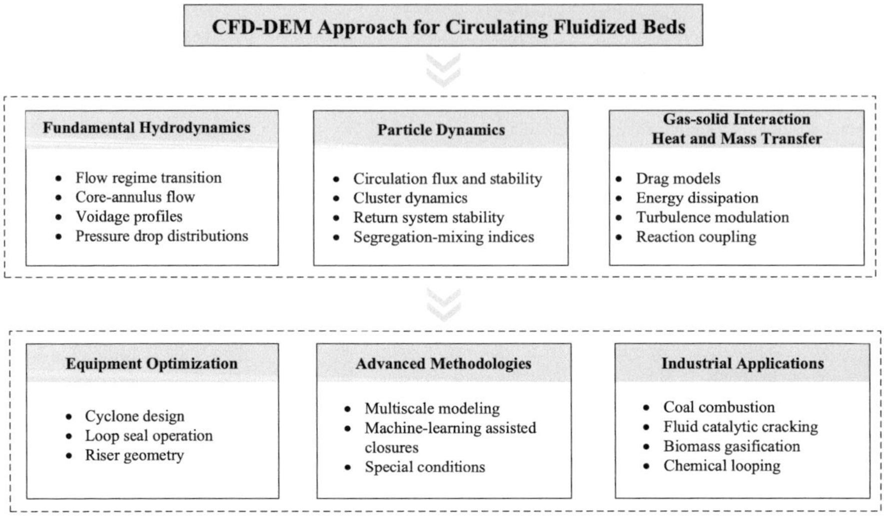

Fig. 9.8 CFD-DEM research framework for CFBs

analysis. For polydisperse systems, researchers examine the competing mechanisms of particle segregation and mixing during continuous circulation.

Gas–solid interaction mechanisms are extensively investigated through coupled simulations. Comparative studies evaluate drag models for their accuracy in predicting cluster-dominated flows characteristic of CFBs. Energy dissipation analysis quantifies the relative contributions of particle collisions, friction and gas-phase viscous effects. The mutual modulation between particle motion and gas turbulence is carefully characterized, revealing how particles affect turbulent eddies while being dispersed by gas turbulence. Heat and mass transfer processes receive significant attention due to their industrial relevance. Local heat transfer coefficients are predicted by coupling energy equations, particularly at riser walls where particle convection dominates. For reactive systems, simulations analyze how solids circulation enhances reactant conversion in applications like coal combustion and gasification. Temperature uniformity is assessed by tracking thermal histories of circulating particles and guiding strategies for thermal management.

Equipment optimization and scale-up studies bridge fundamental research with industrial needs. Parametric investigations evaluate geometric effects like riser aspect ratio and cyclone efficiency on overall system performance. Operating conditions such as gas velocity and solids inventory are optimized to achieve desired density profiles and reaction efficiencies. Scale-up criteria are developed based on similarity principles, preserving key dimensionless numbers across different scales.

Specialized research addresses complex systems and multiphysics phenomena. Studies on cohesive particles such as nanoparticles or wet particles examine interparticle force effects on circulation stability. High-pressure/high-temperature applications like chemical looping combustion are simulated to understand extreme condition behavior. Overall, the CFD-DEM approach offers unique advantages for CFB research, particularly in resolving particle-scale details like cluster microstructure. However, full-scale simulations remain computationally challenging, often requiring advanced techniques like coarse-graining or hybrid modeling approaches.

9.3.2 Example Applications

Shao et al. [5] carried out a detailed numerical study on a chemical looping combustion (CLC) system based on CFB. The study employs a 3D reactive model using the computational particle–fluid dynamics (CPFD) method to investigate the behavior of coal and biomass as fuels, as well as the impact of particle size distributions (PSDs) for both solid fuels and oxygen carriers (OCs). The reaction performance of coal and biomass is compared, highlighting that biomass outperforms coal due to its higher volatile content and lower char content, leading to faster and more efficient conversion to CO_2. Despite shorter residence times for biomass particles, their reaction efficiency is superior, emphasizing the importance of fuel composition in CLC systems. Further analysis explores the effects of PSDs on system performance. For coal, smaller particles (100–200 μm) exhibit better reaction performance despite shorter residence times, as char gasification is the rate-limiting step. In contrast, biomass particles show consistent reaction efficiency across different sizes (300–1100 μm) due to rapid volatile release and low char content. The study also examines the role of OC PSDs, revealing that overly wide distributions (300–800 μm) can compromise system stability by increasing gas leakage risks and reducing separator efficiency. The conversion of OCs is found to depend not only on residence time but also on their trajectories in high-efficiency reaction zones in the fuel reactor, as shown in Fig. 9.9.

Brandt et al. [6] conducted a comprehensive study on the application of a multilevel coarse-graining (MCG) approach in the CFD-DEM framework to simulate a CFB, as shown in Fig. 9.10. The primary objective of the research is to reduce computational costs while maintaining simulation accuracy by strategically adjusting the resolution of the granular phase in different subsystems of the CFB. The conventional coarse-graining (CG) technique is explored to identify a global CG limit of k_{CG}=3 for the CFB, beyond which significant deviations in key parameters like solids circulation rate and holdup occur. The riser subsystem is divided into two configurations: one focusing solely on the riser and another incorporating the L-valve to assess the impact of geometric constraints. The results reveal that the L-valve and riser interaction is critical, requiring a finer resolution to accurately predict solids flow ($k_{CG} = 3$). The cyclone subsystem, while less critical, shows sensitivity to CG ratios in terms of solids holdup and sink velocity. The standpipe subsystem, however,

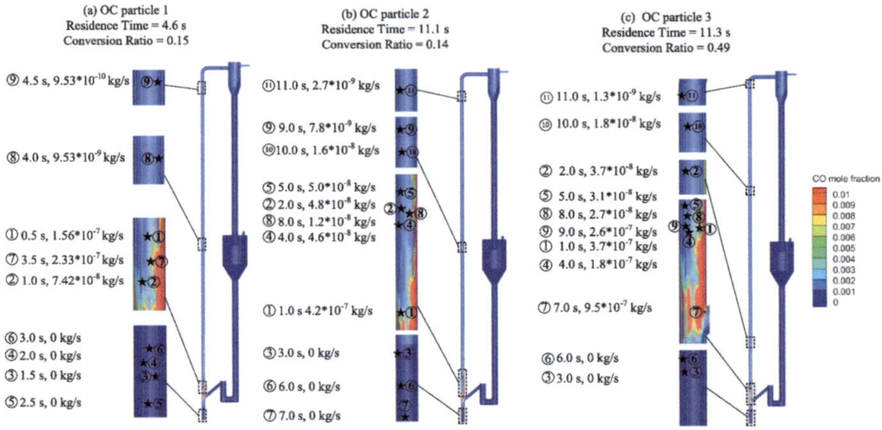

Fig. 9.9 Trajectories and reaction rates over time for three typical OC particles marked in a CLC system based on CFB: **a** OC particle 1 with a residence time of 4.6 s and a final conversion ratio of 0.15; **b** OC particle 2 with a residence time of 11.1 s and a final conversion ratio of 0.14; **c** OC particle 3 with a residence time of 11.3 s and a final conversion ratio of 0.49 [5]

exhibits minimal dependency on CG ratios, allowing for coarser resolutions ($k_{CG} = 6$). Building on these findings, the authors implement the MCG approach, dynamically adjusting grain sizes between subsystems using coarsening and refinement steps. Four spatial configurations of coarsening zones are tested, with the most accurate results achieved when the entire riser and cyclone are resolved with $k_{CG} = 3$ and the standpipe with $k_{CG} = 6$. This configuration reduces the number of modeled entities by 50 times compared to the non-CG simulation while maintaining accuracy. The study highlights the computational efficiency of MCG, demonstrating a 70% faster computation than conventional CG with $k_{CG} = 3$.

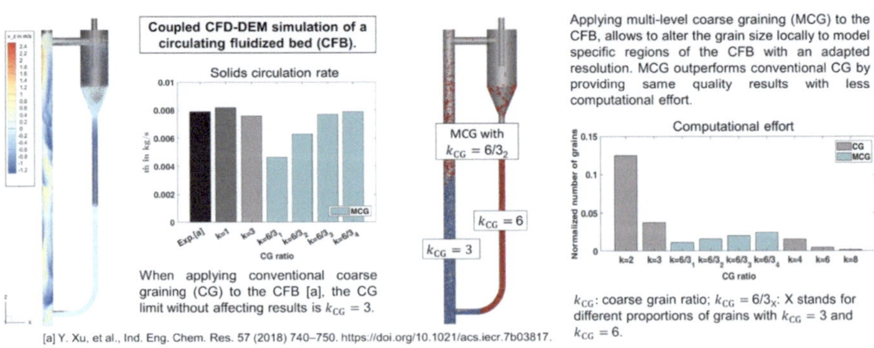

Fig. 9.10 CFD-DEM simulation of a CFB using a particle location-based multilevel coarse-graining approach [6]

9.4 Mixing and Segregation of Particles

9.4.1 Research Focus

Fluidized beds are widely used in industries such as pharmaceuticals, energy and chemical processing due to their excellent mixing capabilities. However, controlling particle mixing and segregation which is critical for process efficiency and product quality remains challenging due to the complex interplay of hydrodynamic and particle-scale phenomena. The coupled CFD-DEM approach has emerged as a powerful tool to unravel these complexities, bridging the gap between individual particle behavior and bulk system performance. This section systematically examines fundamental mechanisms, influencing factors and multicomponent systems, offering a comprehensive framework for understanding and optimizing particle mixing and segregation in fluidized beds. Figure 9.11 displays the CFD-DEM research framework for mixing and segregation of particles in fluidized beds.

(1) Fundamental Mechanisms

The interplay between mixing and segregation establishes dynamic equilibrium states that determine overall bed homogeneity. CFD-DEM studies have revealed intricate mechanisms governing particle mixing and segregation in fluidized beds.

The complex mixing behavior is governed by three primary mechanisms operating at different scales: convective mixing, diffusive mixing and shear mixing. Convective mixing is the large-scale circulation of particles, typically driven by rising gas bubbles or global airflow patterns. This organized motion rapidly transports bulk volumes of material from one region to another, ensuring macroscopic homogenization of the bed. Diffusive mixing is the random and small-scale movement of individual particles caused by collisions and local vibrations. This stochastic motion leads to the gradual blending of adjacent layers, promoting local homogenization at the particle

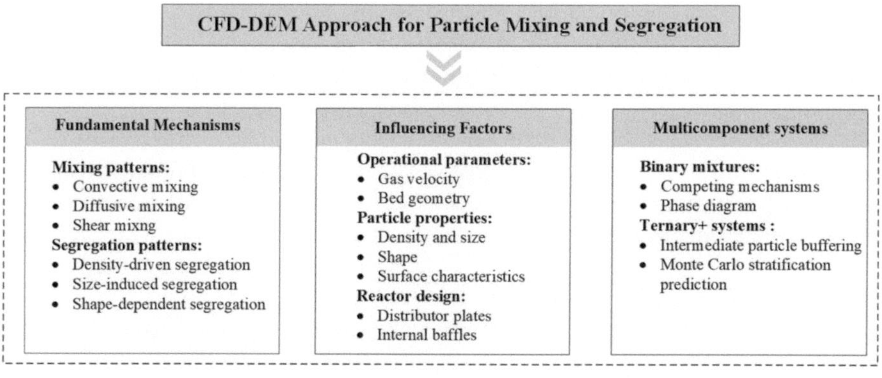

Fig. 9.11 CFD-DEM research framework for mixing and segregation of particles in fluidized beds

scale. Shear-induced mixing occurs due to velocity gradients in the particle bed, where layers of material slide past one another at different speeds. This relative motion causes particles to be exchanged between layers, leading to blending, particularly near walls or around internal obstructions. Quantitative characterization of mixing employs multiple complementary approaches. The Lacey mixing index and relative standard deviation provide statistical measures of composition uniformity across sampling zones. More advanced particle trajectory entropy analysis leverages DEM position data to quantify mixing disorder through information entropy metrics, offering superior resolution for transient mixing processes. Contact number statistics track the frequency of encounters between dissimilar particles, with higher hetero-contacts indicating better mixing. These methods collectively enable comprehensive evaluation of mixing efficiency across various temporal and spatial scales.

As for the segregation mechanism, density-driven segregation follows the Reverse Brazil Nut Effect principle, where denser particles progressively sink through the bed matrix due to greater gravitational potential, while lighter particles accumulate in bubble wakes and rise toward the bed surface. This effect becomes particularly pronounced when density differences exceed 20%. Size-induced segregation operates through two competing mechanisms. Percolation of smaller particles through void spaces between larger particles leads to downward migration, while vibration-induced convection typically pushes larger particles upward, analogous to the classic Brazil Nut Effect. The dominant mechanism depends on particle size ratio and fluidization intensity, with percolation dominating when size ratios exceed 3:1. Particle shape introduces additional segregation complexity. Non-spherical particles like rods or plates exhibit orientation-dependent settling behavior, where alignment with fluid flow reduces drag and promotes upward movement. This anisotropy effect creates stratified layers with distinct orientation distributions. Surface properties further modify segregation through interparticle forces. Electrostatic charges may form stable agglomerates that resist separation, while surface roughness enhances mechanical interlocking that inhibits percolation.

(2) Key Influencing Factors

Operational parameters critically influence mixing and separation behaviors. Gas velocity exhibits a threshold effect: below critical velocities, static stratification prevails, while higher velocities drive turbulent mixing. The bed aspect ratio shapes distribution patterns, with taller beds enhancing axial segregation while improving radial homogeneity. Particle properties induce strong segregation when density differences exceed 20% or size ratios surpass 3:1; however, surface properties such as electrostatic charges or wettability can disrupt conventional behavior. Reactor design elements, including distributor plates and internals, serve as key control mechanisms.

(3) Multicomponent Systems

Multicomponent systems are very common in fluidized bed reactors. Binary mixture studies reveal competing mechanisms when density and size effects oppose each other. Phase diagrams have been developed to predict mixing-segregation transitions, with dynamic tracking showing mixing timescales follow power-law relationships with gas velocity. Ternary systems introduce intermediate particles that buffer extreme segregation, requiring advanced DEM coupled with Monte Carlo sampling for stratification prediction.

9.4.2 Example Applications

Xie et al. [7] conducted a detailed study on the mixing and segregation characteristics of binary column-shaped particles in a liquid–solid fluidized bed (LSFB) using the CFD-DEM coupled with a superquadric model. The research focuses on cylindrical particles with varying aspect ratios (ARs) and density ratios (DRs) to investigate their effects on fluidization dynamics. Results show that heavy particles tend to settle at the bottom of the bed, while light particles rise to the top. The deviation of particle aspect ratio from unity enhances mixing quality and reduces segregation extent compared to approximately spherical particles (AR = 1.0), as shown in Fig. 9.12. Disk-like and rod-like particles exhibit greater rotational motion and require more energy to fluidize due to their irregular shapes and higher collision probabilities. Contact forces are higher for disk-like and rod-like particles due to their irregular shapes, while spherical particles experience the least drag force, inhibiting mixing. Smaller density ratios (DRs) lead to more pronounced segregation, with nearly complete separation observed at DR = 0.60. Larger DR values result in entrainment of heavy particles in light particle regions and vice versa, reducing segregation.

Kazemi et al. [8] investigated the mixing behavior of binary mixtures containing spherical and rod-like particles in a fluidized bed using a coupled CFD-DEM

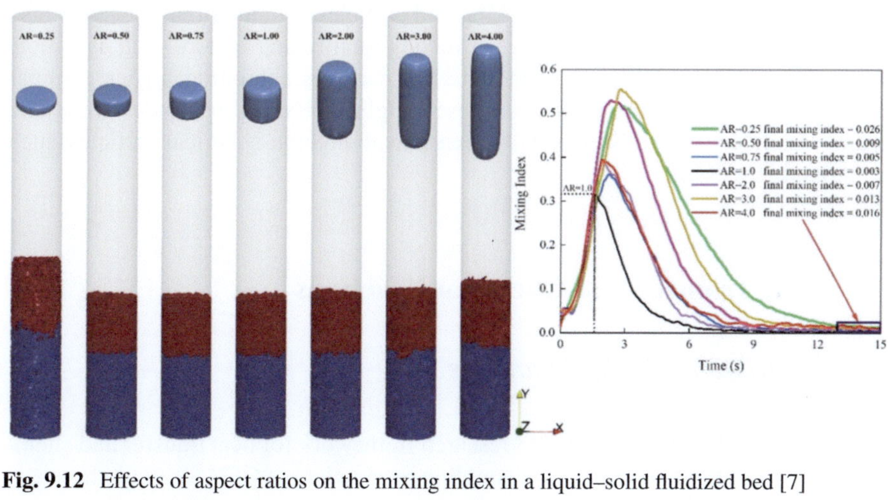

Fig. 9.12 Effects of aspect ratios on the mixing index in a liquid–solid fluidized bed [7]

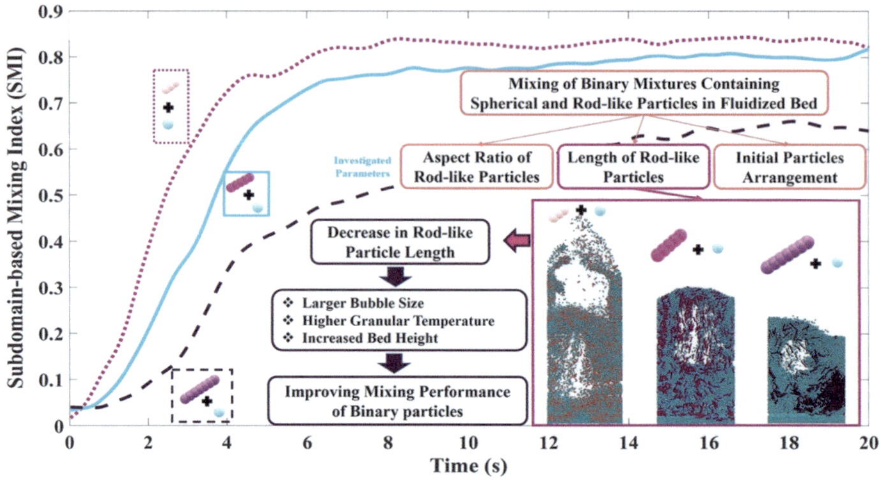

Fig. 9.13 CFD-DEM mixing of rod-like and spherical particles in fluidized beds [8]

approach. The rod-like particles are modeled using the multisphere method, and their motion is tracked using quaternions to link body-fixed and space-fixed coordinate systems. The research focuses on the effects of key parameters, including the volume fraction of spherical particles, the length of rod-like particles, and their aspect ratio, on mixing dynamics, as shown in Fig. 9.13. The subdomain-based mixing index (SMI) is employed to quantitatively evaluate mixing performance, alongside qualitative visual assessments of the fluidized bed. Key findings reveal that increasing the volume fraction of spherical particles enhances mixing due to reduced interlocking and improved fluidization, leading to larger bubbles and higher granular temperatures. Conversely, longer rod-like particles or higher aspect ratios hinder mixing because of increased interlocking and particle accumulation. The study also establishes a direct correlation between mixing quality and factors such as bubble size and granular temperature, with spherical particles exhibiting higher granular temperatures than rod-like particles. These insights are valuable for designing and optimizing industrial fluidized bed processes involving non-spherical particles, such as biomass processing.

9.5 Heat Transfer and Chemical Reactions

9.5.1 Fundamental Mechanisms

Figure 9.14 displays the CFD-DEM research framework for heat transfer and chemical reactions in fluidized beds. The coupling mechanism of CFD for fluid phase and DEM for particle phase has been described in detail in previous chapters.

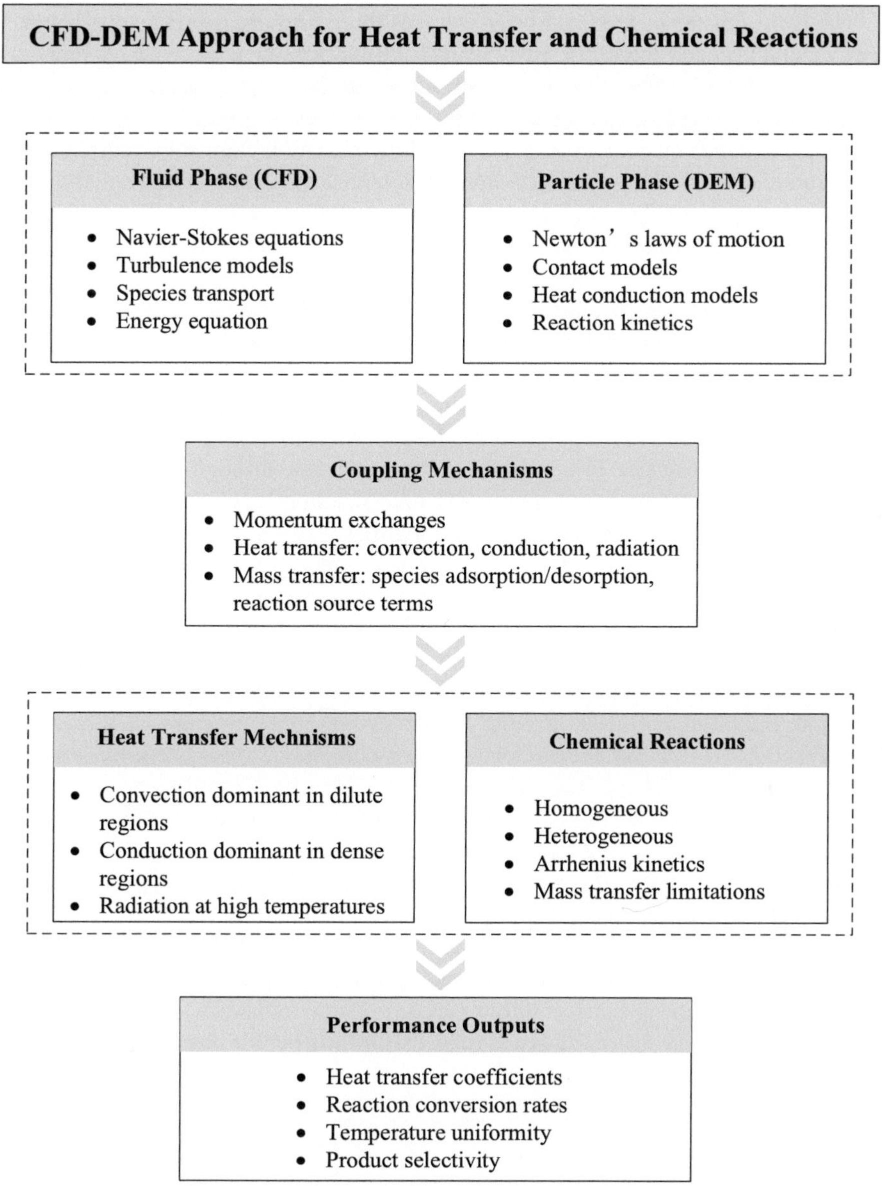

Fig. 9.14 CFD-DEM research framework for heat transfer and chemical reactions in fluidized beds

The heat transfer in fluidized beds involves three primary mechanisms: convection, conduction and radiation. Particle–fluid convection dominates under high gas velocities, where the Nusselt number correlates with Reynolds and Prandtl numbers. Particle–particle conduction becomes significant in dense regions, where the thermal resistance depends on contact area and duration. Radiation gains importance at high temperatures (>800 °C), particularly in regions with large temperature gradients. The dynamic interplay of these mechanisms is influenced by mixing and segregation of particles, bubble formation and disruption, as well as interaction between particles and fluids.

Chemical reactions in fluidized beds can be homogeneous or heterogeneous. Heterogeneous reactions follow adsorption-surface reaction-desorption sequences, with rates governed by Arrhenius kinetics and active site availability. The exothermic or endothermic nature of reactions introduces thermal coupling, where heat release/absorption modifies local temperatures, further accelerating or decelerating reaction rates. Mass transfer limitations such as diffusion through particle pores or gas-phase boundary layers often compete with reaction kinetics, leading to either diffusion-controlled or reaction-controlled regimes.

The efficiency of heat transfer and reactions is highly sensitive to operational and material parameters. Gas velocity enhances convective heat transfer and reactant mixing but may reduce particle contact time. Particle properties such as size and thermal conductivity directly affect conduction, while catalyst loading and porosity dictate reaction accessibility. For BFBs, bubble dynamics introduce nonlinear effects: moderate bubbling improves mixing, but excessive bubbling can isolate particles from reactants. Temperature gradients further complicate the system by inducing thermal stresses or altering reaction equilibrium.

The system's overall performance arises from the nonlinear coupling of mechanisms and key influencing factors. For instance, increasing gas velocity boosts convective heat transfer and reaction rates but may also cause particle elutriation, reducing bed density and conduction. Similarly, exothermic reactions can create localized hot spots, which enhance radiation but risk catalyst deactivation. Optimizing fluidized bed performance thus requires balancing these competing effects. For example, particle size distributions can be tuned to maximize surface area while maintaining adequate heat conduction or by controlling bubble size to ensure uniform reactant distribution.

9.5.2 Example Applications: Combustion/Gasification

Combustion and gasification are essential thermochemical processes in energy conversion and waste treatment, designed to convert carbon-based fuels like coal, biomass or municipal waste into energy and syngas. Combustion fully oxidizes fuel in excess air, producing heat and CO_2, while gasification partially oxidizes it in a controlled oxygen-lean environment, yielding combustible syngas for power generation or chemical synthesis. These processes rely heavily on reaction conditions in the reactors.

The work of Lian et al. [9] presents a comprehensive study on the application of the CFD-DEM approach to investigate biomass combustion in a fluidized bed reactor, as shown in Fig. 9.15, focusing on the effects of cohesive forces between particles due to high inorganic alkali content in biomass. The simulations successfully replicate key phenomena such as particle agglomeration, irregular gas bubbles, rat-holes and gas channels, which are characteristic of cohesive fluidized beds. The results highlight that increasing particle cohesion leads to a continuous decrease in particle kinetic energy and a sudden flattening of the pressure-drop profile at the onset of de-fluidization. Larger cohesive forces are observed with coarser bed particles, while the critical cohesive force at de-fluidization weakens as particle size increases under the same superficial gas velocity. Regions near the bed bottom and sidewalls are identified as more sensitive to hydrodynamic and thermal changes, making them ideal for early detection of agglomeration. The study also examines heat transfer mechanisms, revealing that convection dominates heat transfer for bed particles, whereas reaction heat is predominant for biomass particles. As particle cohesion intensifies, the reaction heat flux of biomass particles decreases due to oxygen-deficient conditions caused by large agglomerates and gas channels.

Du et al. [10] employed a coarse-grained CFD-DEM model to simulate the co-gasification of coal and switchgrass in a BFB reactor, focusing on the effects of inert sand PSD and operating pressure. Results show that increasing the width of the sand PSD has a minimal impact on global reactor hydrodynamics, heat transfer rates and overall gasification performance. However, it causes deteriorated solid mixing at the

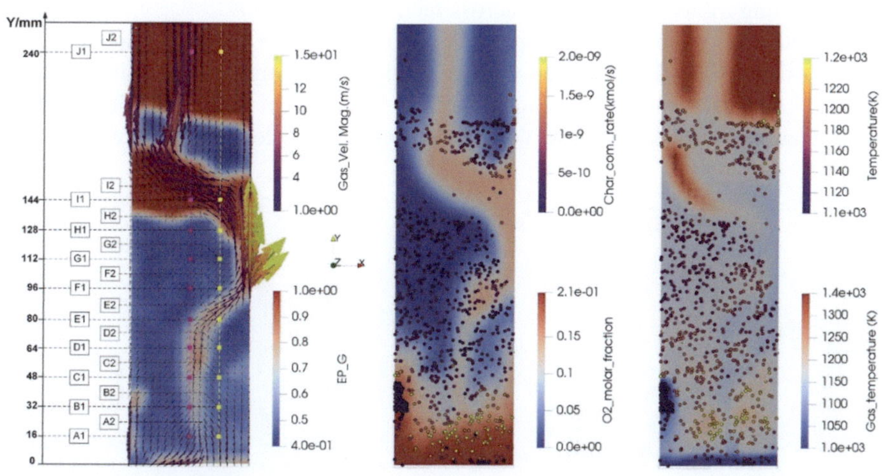

Fig. 9.15 Biomass-particle distribution (colored by combustion rate and temperature) and gas-phase field of voidage, velocity vector, O_2 molar fraction, and temperature at $t = 24.0$ s at the bed-particle size of 1.0 mm [9]

bed bottom, leading to a localized vertical temperature gradient. In contrast, operating pressure has a significant influence. Elevated pressure promotes more uniform flow and enhanced convective heat transfer due to increased gas density. A major finding is that despite causing a substantial decrease in overall bed temperature, higher operating pressure ultimately improves gasification performance. This is because the increased reactant (steam) density accelerates the char conversion reaction, resulting in higher hydrogen production and improved cold gas efficiency.

9.5.3 Example Applications: Pyrolysis/Methanation

Pyrolysis and methanation are designed to convert carbonaceous feedstocks into valuable fuels and chemicals. Pyrolysis thermally decomposes organic materials in the absence of oxygen, producing bio-oil, syngas and solid char, while methanation converts syngas (CO and H_2) into methane through catalytic reactions.

Oyedeji et al. [11] investigated an autothermal pyrolysis of corn stover using CFD-DEM approach. The reactor is a pilot-scale fluidized bed system designed for autothermal pyrolysis, featuring a biomass feed auger, plenum, fluidized bed with silica sand, char cyclones and a bio-oil recovery system. Figure 9.16 displays the effects of equivalence ratio on temperature profile and distribution of fresh biomass and biochar. Results show that increasing equivalence ratio of 0.03–0.09 enhances light gas production (30–41 wt%) but reduces biochar yield (24–17 wt%) due to intensified oxidation. Raising the biomass feed height increases light gas yield and decreases biochar, as higher positions improve oxygen access for oxidation. Wider PSDs slightly increase light gas yield but have minimal impact on bio-oil production, demonstrating robust performance across feedstock variations.

Li et al. [12] conducted a comprehensive investigation of a BFB methanation reactor using CFD-DEM coupled with an Approximate Image Processing Method (AIPM). A local-structure-dependent (LSD) drag model based on the Energy-Minimization Multiscale theory, which resolves mesoscale structures in gas–solid

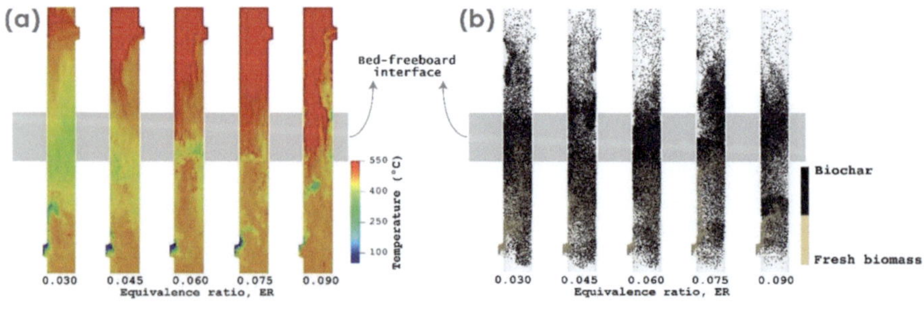

Fig. 9.16 Effects of equivalence ratio on **a** temperature profile and **b** distribution of fresh biomass and biochar [11]

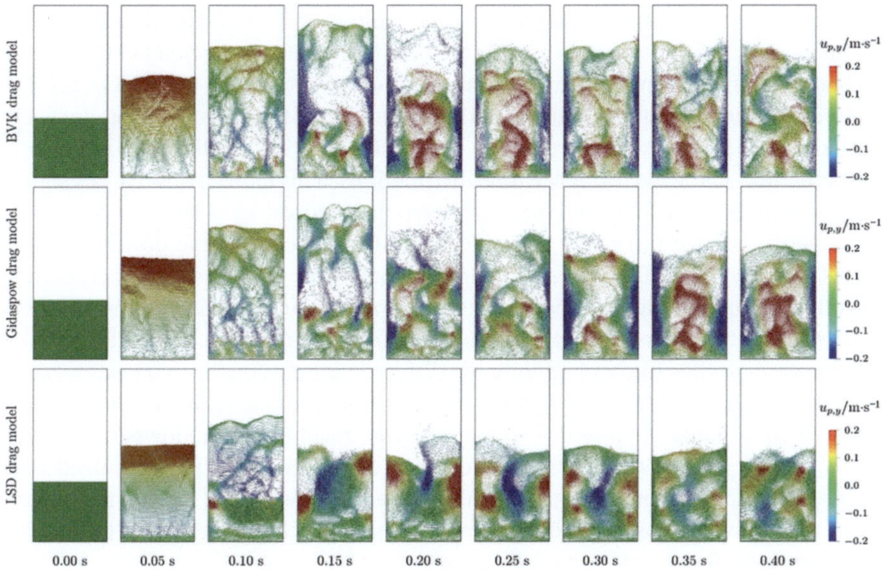

Fig. 9.17 Variation of particle distributions over time using different drag models in a BFB methanation reactor [12]

flows, is integrated. This model outperforms conventional drag models like Gidaspow model and BVK model by more accurately predicting bubble dynamics and bed expansion, as shown in Fig. 9.17. Additionally, modified methanation kinetics are incorporated, accounting for both CO methanation and the water–gas shift reaction, improving the prediction of gas composition and reaction rates. The AIPM successfully captures bubble size, rise velocity and bed expansion height. Results indicate that bubble size increases due to coalescence in the lower bed region but decreases near the freeboard due to breakup.

9.5.4 Example Applications: Fluid Catalytic Cracking

Fluid catalytic cracking (FCC) is a vital process in the petroleum refining industry, designed to break down heavy hydrocarbon molecules into lighter, more valuable products such as gasoline and light olefins. This process relies on a catalyst to facilitate the cracking reactions, typically carried out in riser or downer reactors. Despite its efficiency, FCC faces challenges like catalyst deactivation due to coke deposition and non-uniform flow structures, which can negatively impact product selectivity and yield. Understanding these complexities is crucial for optimizing reactor performance and improving the overall efficiency of the process.

Wu et al. [13] developed a comprehensive CFD-DEM model to simulate gas–solid reacting flows in FCC processes. The model describes heat transfer between particles and gas, instantaneous catalyst deactivation, and gas-phase reactions using lumped kinetics. As shown in Fig. 9.18, the simulations are conducted in 2D domains for both riser and downer reactors, with variations in catalyst-to-oil (CTO) ratios to explore different operating conditions. The results reveal that the riser exhibits severe clustering and backmixing, particularly at high CTO ratios, leading to non-uniform flow structures and wide residence time distributions. This uneven flow causes significant variations in catalyst activity, with particles trapped in clusters experiencing prolonged residence times and faster deactivation. In contrast, the downer demonstrates more uniform flow and near plug-flow behavior, attributed to the co-current downflow alignment with gravity. This uniformity translates into narrower residence time distributions and more consistent catalyst activity throughout the reactor. Temperature profiles further highlight the differences between the two reactor types. In the riser, the gas phase is rapidly heated upon contact with the hot catalyst, followed by gradual cooling due to endothermic reactions. The downer, however, shows weaker heat transfer performance, suggesting the need for optimized inlet designs to ensure proper mixing. From a performance standpoint, the downer outperforms the riser in terms of product selectivity, yielding higher gasoline production and lower coke formation, especially at elevated CTO ratios.

Du et al. [14] developed a comprehensive model integrating the MP-PIC scheme with an eight-lump kinetic model (heavy oil, light oil, gasoline, propane + butane, butylene, propylene, dry gas and coke) to describe the catalytic cracking reactions. An industrial-scale two-stage riser catalytic cracking for maximizing propylene riser reactor is simulated, which features a diameter-enlarged section to enhance gas–solid contact and reaction efficiency, as shown in Fig. 9.19. The reactor is divided into three sections: pre-lifting, diameter-enlarged and conveying. The results reveal distinct regions of product formation, with propylene primarily generated in the diameter-enlarged section due to high catalyst-to-oil ratios and optimal temperatures. Heavy oil cracking occurs predominantly in the upper sections, producing lighter hydrocarbons. Particles exhibit chaotic motion in the diameter-enlarged section, with some entrapped for extended periods, leading to longer residence times and higher coke deposition. The diameter-enlarged section shows a wide residence time distribution due to intense turbulence, while the conveying section has a narrower distribution. The mean residence time in the reactor is 4.22 s, with some particles remaining trapped for over 8 s. Particles with longer residence times accumulate more coke, reducing their activity. However, over 90% of particles maintain a relative activity above 0.6 at the reactor outlet, indicating efficient operation.

9.6 Challenges

The application of CFD-DEM approaches to the fluidized bed systems faces several challenges that require innovative solutions to advance both fundamental understanding and industrial applications. These challenges primarily include two critical areas: multiphysics coupling phenomena and ultrafine particle behavior, both

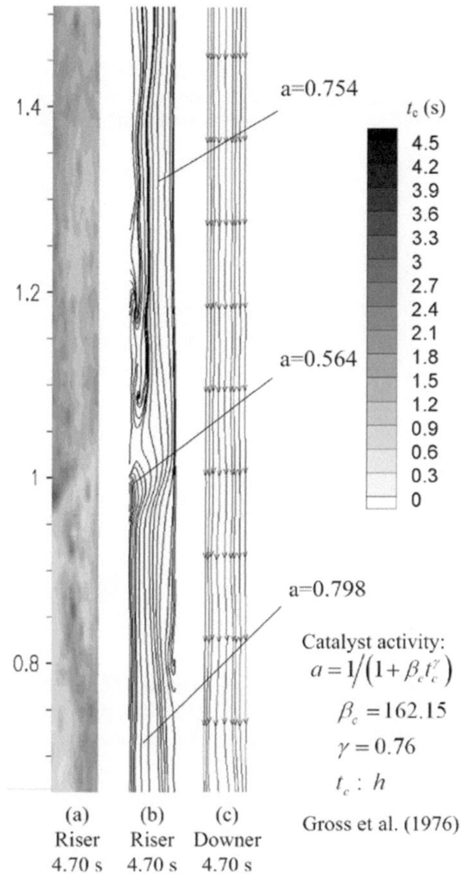

Fig. 9.18 Spatial distributions of particle residence time, activity of catalyst, and the streamlines: **a** cell-averaged particle residence time; **b** streamlines in the riser; and **c** streamlines in the downer [13]

presenting unique complexities that demand targeted research efforts. For multi-physics coupling, the intricate thermo-fluid-chemical interactions in reactive systems create nonlinear feedback loops among temperature fields, hydrodynamic behavior, and chemical transformations. The modeling of ultrafine particle systems represents another major frontier. When dealing with particles smaller than 10 microns, conventional DEM approaches break down as interparticle forces like van der Waals interactions and electrostatic effects become dominant over inertial forces. Recent advances have developed some novel models to account for these molecular-scale interactions. However, these improvements reveal new complexities, such as the "pseudo-mixing" phenomenon where ultrafine particles form stable agglomerates that appear well-mixed but actually exhibit significantly reduced reactivity. The computational cost is also very challenging, as simulating realistic quantities of ultrafine particles at full resolution remains prohibitively expensive, driving the development of innovative coarse-graining techniques that balance accuracy with computational feasibility.

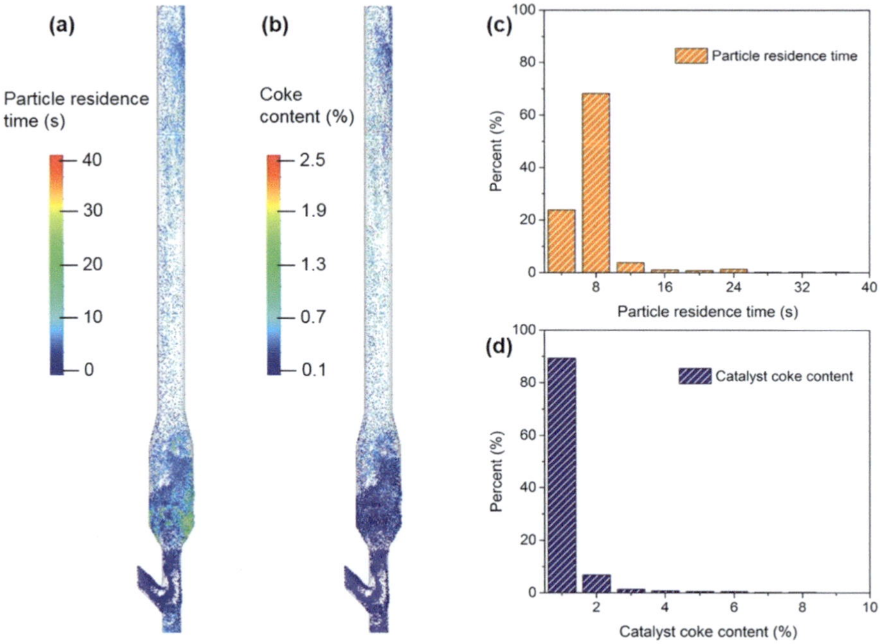

Fig. 9.19 Instantaneous distributions of **a** the particle residence time and **b** the catalyst coke content in the multiregime riser reactor at 40 s; probability distributions of **c** the particle residence time and **d** the catalyst coke content from the riser reactor outlet [14]

References

1. X. Zhu et al., Impact of pressure on gas-solid hydrodynamics of Geldart B and D particles in a pressurized bubbling fluidized bed: a CFD-DEM study. Particuol. **96**, 328–340 (2025). https://doi.org/10.1016/j.partic.2024.11.016
2. K. Korkerd, Z. Zhou, R. Zou, P. Piumsomboon, B. Chalermsinsuwan, Effect of immersed tubes configurations on mixing and heat transfer of mixed biomass and silica sand in a bubbling fluidized bed using CFD-DEM and statistical experimental design analysis. Pow. Technol. **437** (2024). https://doi.org/10.1016/j.powtec.2024.119542
3. Z. Luo, Y. Yue, S. Wang, Y. Shen, CFD-DEM study of spout deflection behavior of cohesive particles in a spout fluidized bed. Powd. Technol. **427** (2023). https://doi.org/10.1016/j.powtec.2023.118758
4. Z. Luo, S. Wang, Y. Shen, Super-quadric CFD-DEM study of spout deflection behaviour of non-spherical particles in a spout fluidized bed. Pow. Technol. **433** (2024). https://doi.org/10.1016/j.powtec.2023.119240.
5. Y. Shao, Y. Hu, X. Wang, B. Jin, R.K. Agarwal, Numerical study of a chemical looping combustion system: effects of fuel type and particle size distributions on the operating performance. Chem. Eng. J. **500** (2024). https://doi.org/10.1016/j.cej.2024.156476
6. V. Brandt, J. Grabowski, N. Jurtz, M. Kraume, H. Kruggel-Emden, CFD-DEM simulation of a circulating fluidized bed using a particle location based multi-level coarse-graining approach. Powder Technol. **460** (2025). https://doi.org/10.1016/j.powtec.2025.121075

7. L. Xie, S. Wang, B. Shao, X. Chen, N. Ding, Y. Ma, CFD-DEM study on mixing and segregation characteristics for binary column-shape particles in a liquid–solid fluidized bed. Adv. Pow. Technol. **36**(2) (2025). https://doi.org/10.1016/j.apt.2025.104788
8. S. Kazemi, R. Zarghami, N. Mostoufi, R. Sotudeh-Gharebagh, J. Chaouki, CFD-DEM mixing of rod-like and spherical particles in fluidized beds. Pow. Technol. **442** (2024). https://doi.org/10.1016/j.powtec.2024.119847
9. G. Lian, W. Zhong, Coupling CFD-DEM with cohesive force and chemical reaction sub-models for biomass combustion in a fluidized bed. Fuel **350** (2023). https://doi.org/10.1016/j.fuel.2023.128858
10. S. Du, J. Wang, Y. Yu, Q. Zhou, Coarse-grained CFD-DEM simulation of coal and biomass co-gasification process in a fluidized bed reactor: effects of particle size distribution and operating pressure. Renew. Energy **202**, 483–498 (2023). https://doi.org/10.1016/j.renene.2022.11.073
11. O.A. Oyedeji et al., CFD–DEM modeling of autothermal pyrolysis of corn stover with a coupled particle- and reactor-scale framework. Chem. Eng. J. **446** (2022). https://doi.org/10.1016/j.cej.2022.136920
12. J. Li, R.K. Agarwal, L. Zhou, B. Yang, Investigation of a bubbling fluidized bed methanation reactor by using CFD-DEM and approximate image processing method. Chem. Eng. Sci. **207**, 1107–1120 (2019). https://doi.org/10.1016/j.ces.2019.07.016
13. C. Wu, Y. Cheng, Y. Ding, Y. Jin, CFD-DEM simulation of gas-solid reacting flows in fluid catalytic cracking (FCC) process. Chem. Eng. Sci. **65**(1), 542–549 (2010). https://doi.org/10.1016/j.ces.2009.06.026
14. Y. Du, X. Chen, S. Li, A. S. Berrouk, W. Ren, C. Yang, Revisiting a large-scale FCC riser reactor with a particle-scale model. Chem. Eng. Sci. **249** (2022). https://doi.org/10.1016/j.ces.2021.117300

Chapter 10
Applications to Magnetized and Electrostatic Fluidized Beds

10.1 Magnetized Fluidized Beds

A magnetized fluidized bed (MFB) is an advanced reactor configuration that integrates a magnetic field with a conventional fluidized bed to enhance control over particle behavior and improve process efficiency. Introduced in the 1960s, MFBs utilize magnetically susceptible particles such as ferromagnetic, paramagnetic or diamagnetic materials, which respond to an externally applied magnetic field [1]. The magnetic susceptibility of these particles determines their interaction with the field, influencing bed hydrodynamics, stability and performance. By adjusting the magnetic field intensity and orientation, operators can manipulate particle movement, suppress bubbling and optimize gas–solid contact, addressing common challenges in traditional fluidized beds like particle agglomeration and uneven distribution. This technology has evolved significantly, finding applications in catalytic reactions, adsorption, filtration and separation processes, driven by its ability to intensify heat and mass transfer while maintaining low pressure drops.

10.1.1 Classifications

MFBs can be systematically classified based on their operational modes, magnetic field orientations, fluidization regimes and the types of magnetic fields applied, as shown in Fig. 10.1. Each classification criterion plays a crucial role in determining the hydrodynamic behavior, stability and efficiency of the bed.

(1) Operational Modes

The mode of operation in MFBs refers to the sequence in which the magnetic field and fluidizing gas flow are introduced, significantly influencing bed stability and particle dynamics. As shown in Fig. 10.2a, in the magnetic first mode, the magnetic

R. K. Agarwal and Y. Shao, *Theory and Application of Discrete Element Method to Simulation of Multiphase Flows*, https://doi.org/10.1007/978-3-032-11157-9_10

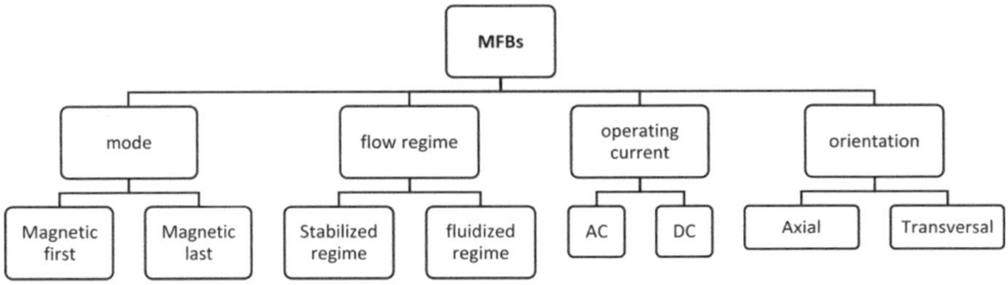

Fig. 10.1 Illustration of major classifications of MFBs [2]

field is applied before the gas flow begins. This approach stabilizes the bed structure in advance, creating a rigid network of magnetically aligned particles that resist bubbling and channeling when fluidization starts. This mode is particularly useful in processes requiring a homogeneous particle distribution, such as catalytic reactions or adsorption. Conversely, the magnetic last mode involves initiating gas flow first, followed by the application of the magnetic field, as shown in Fig. 10.2b. This method allows the bed to initially behave like a conventional fluidized bed before the magnetic field restructures the particle arrangement. The transition can lead to unique hydrodynamic phenomena, such as the formation of anisotropic particle chains or enhanced segregation in mixed-particle systems. Additionally, the on–off magnetization mode, where the magnetic field is periodically switched on and off, has been explored to improve heat transfer and mitigate temperature gradients, though its applications remain less studied compared to the other modes.

(2) Magnetic Field Orientations

The orientation of the magnetic field relative to the fluidizing gas flow is another critical classification parameter, as shown in Fig. 10.2c and d. In axial MFBs, the magnetic field is applied parallel to the direction of gas flow. This alignment causes particles to form linear chains along the field lines, which enhances bed stability and reduces bubble formation. Axial fields are commonly used in magnetically stabilized bed for applications like filtration and catalytic processes, where uniform particle distribution is essential. In contrast, transversal MFBs employ a magnetic field perpendicular to the gas flow. This configuration induces different particle structuring behaviors, such as the formation of columnar aggregates that can improve mixing or segregation depending on the process requirements. Transversal fields are particularly effective in separation processes, such as coal beneficiation or mineral sorting, where magnetic and non-magnetic particles must be efficiently segregated. The choice between axial and transversal fields depends on the desired balance between stabilization and dynamic particle movement.

(3) Fluidization Regime

MFBs exhibit distinct fluidization regimes depending on the gas velocity and magnetic field strength. The magnetically stabilized regime occurs when the gas velocity is between the minimum fluidization velocity and the minimum bubbling

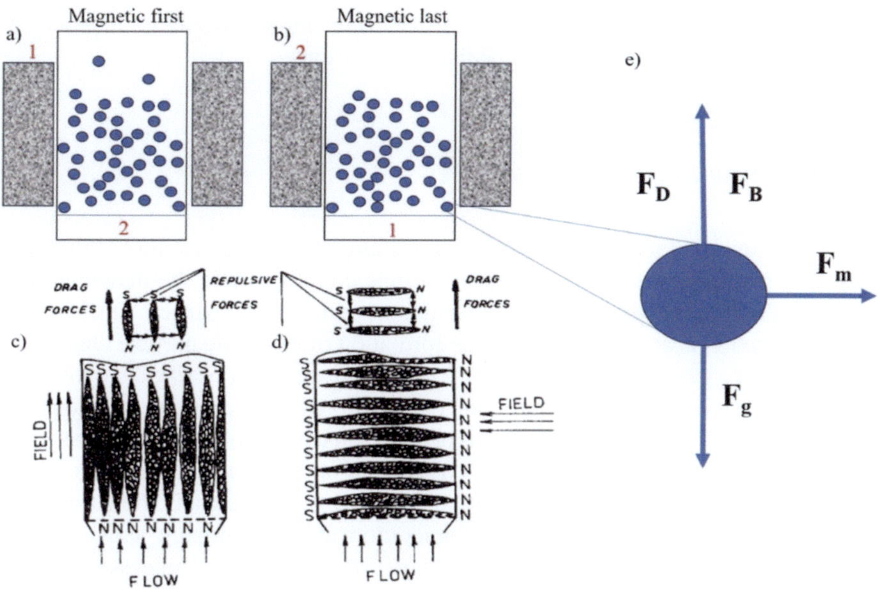

Fig. 10.2 Illustration of **a** MFB in magnetic first mode and **b** MFB in magnetic last mode; representation of particle arrangement as strings in **c** axial, **d** transversal field and **e** forces acting on a single particle in an MFB [2]

velocity. In this regime, the bed expands uniformly without bubbling, resembling the behavior of Geldart-A powders in conventional fluidization. This stabilization is ideal for processes requiring prolonged gas–solid contact, such as adsorption or chemical synthesis. At higher gas velocities, the magnetically fluidized regime emerges, characterized by bubbling and particle mixing. While the magnetic field still influences particle dynamics, its stabilizing effect is partially overcome by the increased gas flow. This regime is useful for applications demanding vigorous mixing, such as catalytic reactions with high heat and mass transfer requirements. Beyond a critical magnetic field strength, the bed may enter a magnetically frozen or condensed regime, where particle movement is severely restricted, forming a rigid, anisotropic structure. This state is typically avoided in continuous processes but can be exploited in specialized applications like filtration or particle immobilization.

(4) Magnetic Field Type

MFBs can also be categorized based on the type of magnetic field applied, including direct current (DC) and alternating current (AC). DC fields provide a constant magnetic intensity, leading to steady particle alignment and predictable bed behavior. They are widely used in industrial applications due to their simplicity and effectiveness in stabilization. AC fields, on the other hand, introduce time-varying magnetic forces that can induce oscillatory particle motion. While less common, AC fields have been explored for their potential to disrupt agglomerates or enhance mixing in cohesive powders. Additionally, rotating magnetic fields have been studied for their ability to create dynamic particle movement, offering unique advantages in processes like nanoparticle fluidization or biomedical applications.

10.1.2 Functions and Advantages

MFBs offer unique functionalities that address limitations of conventional fluidized beds. One primary function is enhanced particle control, where the magnetic field stabilizes the bed, suppresses bubble formation and reduces particle entrainment. This stabilization is achieved through the formation of magnetic particle chains that resist disruptive gas flows. Such control is critical in applications like catalytic reactions, where uniform particle distribution ensures consistent reaction conditions and improved yields. Another key function is process intensification. MFBs improve heat and mass transfer rates by optimizing bed porosity and gas–solid contact areas. For instance, in methane reforming or ammonia synthesis, the magnetic field enhances catalyst efficiency by minimizing agglomeration and maximizing active surface exposure. The technology also excels in separation and filtration processes, such as coal beneficiation or flue gas desulfurization, where magnetic stabilization enables selective particle separation based on density or magnetic properties. Furthermore, MFBs enable diverse industrial applications due to their operational flexibility. In chemical engineering, they are used for hydrogen production, olefin-paraffin separation and nanoparticle fluidization. In environmental engineering, MFBs efficiently remove pollutants like sulfur dioxide or particulate matter from gases. Their adaptability is further demonstrated in biomedical and material science applications, such as yeast fermentation or carbon nanotube growth, where precise particle control is important.

10.1.3 Example Applications

Wang et al. [3] investigated the segregation behaviors of binary mixtures in magnetic fluidized beds using a CFD-DEM approach enhanced with a magnetic field force model. The study focuses on how magnetic field intensity, gradient and superficial gas velocity influence particle segregation. The magnetic field force is described in detail, consisting of two components: the external magnetic field gradient force F_{me} and the interparticle magnetic field force F_{mi}. The external force F_{me} arises from the gradient of the magnetic field and acts opposite to gravity, calculated as $F_{\mathrm{me}} = \frac{V_p \chi_f B_0(\beta_0 + k_0 h)}{\mu_0}$, where V_p is particle volume, χ_f is susceptibility, B_0 is magnetic induction intensity, k_0 is the gradient and μ_0 is vacuum permeability. The interparticle force F_{mi} becomes significant when particles are within twice their diameter and includes radial and tangential components derived from dipole interactions and are dependent on particle alignment and distance. The main findings reveal that

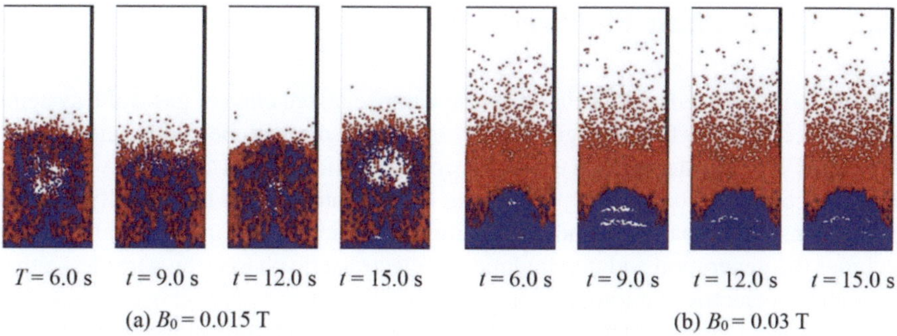

Fig. 10.3 Instantaneous particle distributions at different magnetic field intensities [3]

moderate magnetic field intensities enhance segregation by lifting magnetic particles to the upper bed region, while excessive intensities cause agglomeration, reducing separation efficiency, as shown in Fig. 10.3. The magnetic field gradient also plays a critical role. A higher gradient improves segregation, but beyond an optimal range, it leads to particle aggregation at lower bed heights, diminishing separation quality. A dimensionless parameter, the magnetic-gravity-ratio number ($F_{\text{mec}} = |F_{\text{me}}|/|G|$), is introduced to quantify the balance between magnetic and gravitational forces, with effective segregation occurring when the magnetic-gravity-ratio number ranges between 0.29 and 0.38.

Wang et al. [4] presented a detailed study on the simulation of particle flow behavior in a liquid–solid fluidized bed under a uniform axial magnetic field using a CFD-DEM approach. The effects of magnetic field strength on particle distribution, velocity, concentration and granular temperature in the bed are investigated. A two-dimensional CFD-DEM model is used to simulate the interactions between liquid and solid phases, incorporating forces such as drag, virtual mass, lubrication, contact and magnetic forces. The magnetic field is applied in the "magnetization LAST" mode, where the bed is fluidized before magnetization. The simulation results reveal that at low magnetic flux densities, particles exhibit fluidization with small chain formations, while moderate fields lead to the formation of straight-chain aggregates aligned with the magnetic field direction. These chains enhance particle fluctuations and granular temperature. However, at high magnetic flux densities, particles form rigid, frozen structures, leading to defluidization. The simulation results also observe segregation between magnetic and non-magnetic particles, with magnetic particles settling at the bottom due to increased weight from chain formation. The granular temperature of particles, which reflects their kinetic energy, initially increases with magnetic flux density, peaks at moderate fields and then declines as chains immobilize particles.

10.2 Electrostatic Fluidized Beds

Electrostatic fluidized beds (EFBs) represent a specialized class of gas–solid systems where electrostatic forces significantly influence particle behavior and system dynamics. These systems build upon conventional fluidized bed technology by incorporating controlled or naturally occurring electrostatic effects that modify particle interactions, flow patterns and separation processes. The fundamental concept includes the generation and manipulation of electrostatic charges on particles, either through triboelectric charging during particle collisions or by applying external electric fields. This dual mechanism of charge generation creates unique opportunities for process intensification while also presenting challenges in system control and stability.

10.2.1 Classifications

The classification of electrostatic fluidized beds can be approached from multiple perspectives. Based on charge generation mechanisms, it can be classified as naturally charged systems where triboelectrification dominates and externally controlled systems using applied electric fields. Another classification considers operational objectives: coating systems leverage particle charging for uniform deposition; separation systems exploit differential charging for material purification; stabilization systems use electrostatic effects to enhance fluidization quality. Process-scale classifications further divide these systems into laboratory-scale experimental units, pilot-scale demonstration systems and full-scale industrial implementations, each with distinct design considerations and operational challenges.

The applications for electrostatic fluidized beds can be seen in various industries. In materials processing, electrostatic fluidized beds enable precise powder coating applications by facilitating superior surface adhesion through particle charging. Environmental engineering utilizes electrostatic fluidized beds for efficient separation of mixed plastic wastes in recycling operations. The pharmaceutical industry applies electrostatic fluidization principles to improve powder handling and formulation processes. The energy sector utilizes these systems for advanced coal beneficiation and biomass processing applications. More detailed summary of application of electrostatic fluidized beds can be found in the review [5].

10.2.2 Example Applications

CFD-DEM simulations on electrostatic fluidized beds can capture essential phenomena including charge accumulation, particle trajectories and field distributions while accounting for complex hydrodynamic behaviors. Current research mainly focuses on several critical areas. Bubble dynamics under electrostatic influence remains a primary investigation topic, particularly how electric fields modify bubble formation, growth and bursting patterns. Particle segregation studies examine charge-based separation mechanisms for mixed material systems, with applications in recycling and mineral processing. Another research focus involves stability analysis, exploring how electrostatic forces affect bed expansion, particle elutriation and agglomeration tendencies.

Liu et al. [6] investigated the effects of a direct current electric field on gas–solid fluidized beds using the CFD-DEM approach. The study focuses on how the electric field influences mesoscale structures like bubbles and clusters. A 2D numerical model is established, incorporating both local-field effects and multipolar interactions in the calculations of electric-field-induced forces. The forces are modeled using a revised point-dipole approximation, which accounts for the polarization of particles and their interactions under the electric field. The electric field induces dipole moments in the particles, leading to attractive or repulsive forces depending on the alignment of the particles relative to the field, as shown in Fig. 10.4. The study highlights that the local-field effect, which considers the influence of neighboring particles on polarization, is about twice as significant as multipolar interactions. As the electric field strength increases, the formation of particle chains suppresses vortex motions and reduces bubble sizes, ultimately leading to more homogeneous distributions of particle concentration and velocity, as shown in Fig. 10.5. The chains exhibit a unique "braided" structure, where particles align obliquely to avoid repulsive forces in the horizontal direction. The results also demonstrate that higher electric field strengths enhance particle interactions, increase contact forces and reduce drag forces for particles in chains.

Further, Kazemi et al. [7] expanded the investigation to more industrially relevant rodlike particles using a sophisticated multisphere approach with quaternion-based

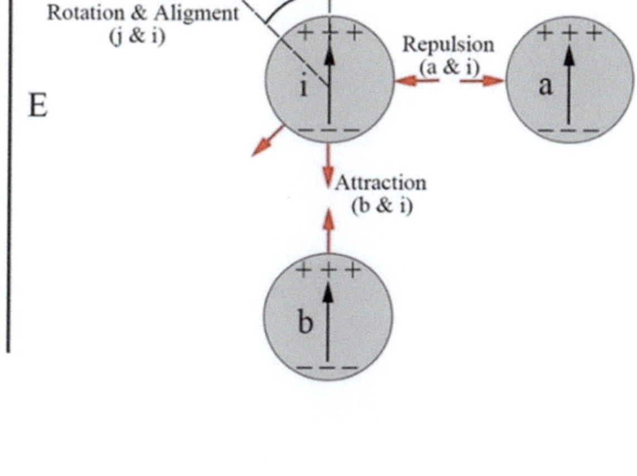

Fig. 10.4 Polarization force model among particles in a direct current electric field [6]

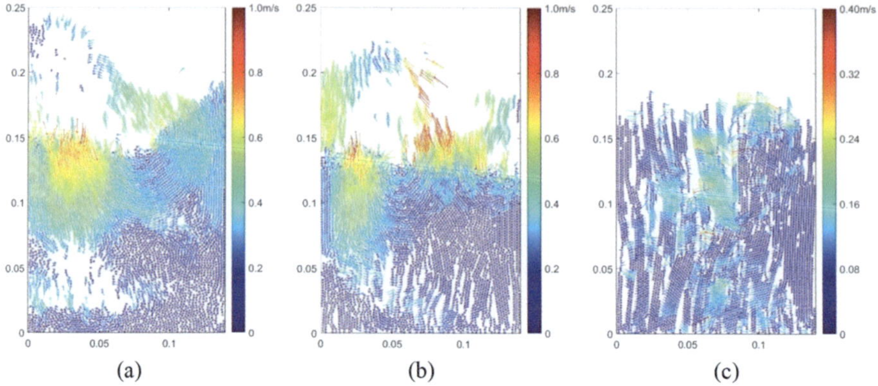

Fig. 10.5 Distribution of particle velocity vector at various field strengths at $t = 4$ s: **a** $E = 0.75$ kV/mm, **b** $E = 1.5$ kV/mm, **c** $E = 3$ kV/mm [6]

orientation tracking. The work demonstrates that particle shape dramatically affects response to electric fields, with rodlike particles showing stronger alignment tendencies and forming complex braided structures rather than simple chains, as shown in Fig. 10.6. Similar to the work of Liu et al. [6], the key findings also show that electric fields can effectively reduce bubble size and homogenize particle distribution in fluidized beds. By suppressing bubble growth and preventing coalescence, electric fields enhance gas–solid contact, which is beneficial for industrial processes requiring efficient mass and heat transfer. Additionally, the fields influence particle orientation, particularly for non-spherical particles like rods, which tend to align with the field direction. Smaller particles exhibit greater alignment due to their lower inertia, while larger particles resist reorientation. The electric field's impact on bed porosity is also notable, as it compresses the emulsion phase and reduces voidage fluctuations, leading to more stable fluidization. By controlling particle organization and bubble behavior, electric fields can improve reactor performance without mechanical modifications. However, excessive field strengths may lead to defluidization or unwanted particle agglomeration, necessitating careful optimization.

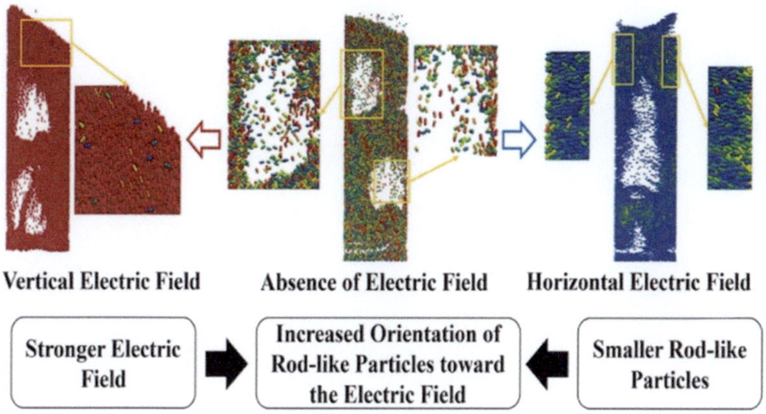

Fig. 10.6 Effect of external electric field on fluidization of rodlike particles [7]

References

1. F. Filippov, The effect of a magnetic field on a ferromagnetic particle suspension bed. Prik. Magnit. Lat. SSR **12**, 215 (1960)
2. B. Deivendran et al., *Gas-Solid Magnetically Fluidized Beds: A Review*. (Elsevier, 2025). https://doi.org/10.1016/j.powtec.2025.120615
3. B. Wang, T. Tang, S. Yan, Y. He, Magnetic segregation behaviors of a binary mixture in fluidized beds. Powder Technol. **397**, 7031 (2022). https://doi.org/10.1016/j.powtec.2021.117031
4. S. Wang, Z. Sun, X. Li, J. Gao, X. Lan, Q. Dong, Simulation of flow behavior of particles in liquid-solid fluidized bed with uniform magnetic field. Powder Technol. **237**, 314–325 (2013). https://doi.org/10.1016/j.powtec.2012.12.013
5. F. Fotovat, X.T. Bi, J.R. Grace, *Electrostatics in Gas-Solid Fluidized Beds: A Review*. (Elsevier, 2017). https://doi.org/10.1016/j.ces.2017.08.001
6. G. Liu et al., CFD-DEM study of the effects of direct current electric field on gas-solid fluidization. Powder Technol. **362**, 416–427 (2020). https://doi.org/10.1016/j.powtec.2019.11.096
7. S. Kazemi, H. Aali, R. Saghafian-Larijani, R. Zarghami, H. Liu, N. Mostoufi, Effect of external electric field on fluidization of rodlike particles using CFD-DEM. Ind. Eng. Chem. Res. **63**(48), 20937–20949 (2024). https://doi.org/10.1021/acs.iecr.4c02474

Chapter 11
Applications to Dense Granular Systems

11.1 Particle Packing

Particle packing refers to the spatial arrangement and distribution of discrete particles in a confined volume, influencing properties such as density, porosity and mechanical stability. It has widespread applications across various scientific and engineering fields. In geotechnical engineering, it determines the strength and permeability of soils, critical for foundations and slope stability. Industrial processes rely on optimized packing for efficient powder compaction in pharmaceuticals, ceramics and battery materials. In construction, dense particle arrangements enhance the durability of concrete and asphalt by minimizing voids. Environmental engineering uses packing principles in filtration systems and landfill design to improve efficiency and safety. Additionally, materials science leverages controlled packing to develop advanced composites, catalysts and coatings with tailored properties.

The packing structure of particulate systems mainly depends on two factors: particle-level characteristics and system-level conditions. At the particle level, geometric properties play a crucial role, where spherical particles generally achieve higher packing densities than non-spherical ones due to their ability to rearrange more efficiently. Particle size distribution also significantly influences packing, with bimodal or polydisperse systems often exhibiting higher densities than monodisperse systems as smaller particles fill voids between larger ones. Surface properties, particularly interparticle friction and cohesive forces, further affect packing outcomes. High friction restricts particle mobility, leading to looser configurations, while adhesive forces from moisture, van der Waals interactions or electrostatic charges promote clustering and typically increase porosity, especially in fine powders where these forces dominate over gravity. System-level conditions introduce additional complexity to packing behavior. Mechanical methods like static compression or dynamic vibration can obviously increase packing density by facilitating particle rearrangement, though vibration parameters must be carefully controlled as excessive energy input may cause segregation or fluidization. Flow conditions during particle

R. K. Agarwal and Y. Shao, *Theory and Application of Discrete Element Method to Simulation of Multiphase Flows*, https://doi.org/10.1007/978-3-032-11157-9_11

deposition, including pouring height and rate, influence initial packing structure, with gentler deposition generally producing looser packing. Container geometry and boundary effects also play important roles, where wall interactions often create density gradients in confined systems.

Based on structural order and formation mechanisms, packing structures can be classified into several types including random packing, ordered packing, non-spherical particle packing and dynamic process-dependent packing. Comprehensive understanding of particle packing provides a valuable foundation for optimizing industrial processes where precise control over packing density and arrangement is critical.

11.1.1 Random Packing

Random packing refers to the disordered arrangement of particles in a granular material, where no long-range order exists in their spatial distribution. This type of packing is ubiquitous in nature and industrial applications. Random packing mainly includes random loose packing (RLP) and random close packing (RCP). RLP represents the least dense configuration achievable under gentle conditions, such as pouring particles without external compaction. With a porosity typically ranging from 35 to 45% [1], RLP structures are highly susceptible to rearrangement under shear or vibration. RCP describes the densest possible disordered arrangement of particles, obtained through vibration, compression or other densification methods. It exhibits lower porosity of around 25–35% and higher mechanical rigidity compared to RLP [2].

Washino et al. [3] performed DEM simulations to investigate the loose packing behaviors of fine and cohesive particles, focusing on the roles of interparticle forces and cluster formation during free fall. The results demonstrate that achieving realistic low packing fractions requires both proper interparticle force modeling and controlled cluster formation during free fall. While non-bonded van der Waals forces are essential for accurate simulations, adjusting the surface energy density in JKR adhesion models can produce equivalent results. The study highlights the critical role of particle insertion methods. Stream insertion with velocity fluctuations which is optimal at 0.5 m/s enables natural cluster formation and lower packing densities compared to at-once insertion, as shown in Fig. 11.1. The work also shows how particle shape affects packing, with cubic particles forming dense face-aligned structures at low cohesion but porous arrangements at high cohesion.

11.1.2 Ordered Packing

Ordered packing represents another major category and refers to the highly structured arrangement of particles in a periodic, repeating pattern, characterized by

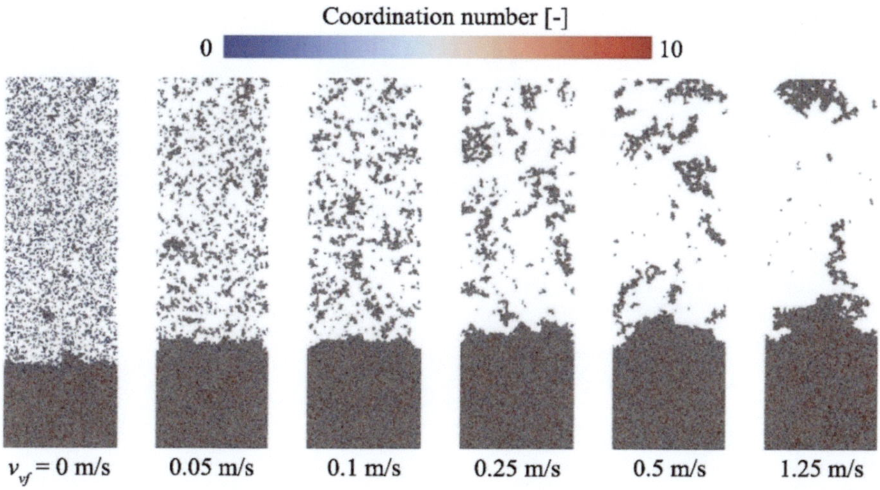

Fig. 11.1 Snapshots of clusters with different vertical velocity fluctuation values at time 70 ms [3]

long-range symmetry and uniformity. Unlike random packing, where particles are distributed irregularly, ordered packing maximizes spatial efficiency, often achieving higher packing densities. Common examples include crystalline structures like face-centered cubic (FCC) and hexagonal close-packed (HCP) arrangements, where spherical particles occupy up to 0.74 of the available volume [4], which is the theoretical maximum for monodisperse spheres. While rarely occurring in natural granular systems due to entropy-driven disorder, ordered packings serve as important reference cases in DEM studies. Simulations of FCC and HCP structures are typically initialized by placing particles in predefined lattice positions, providing benchmarks for ideal packing scenarios. Other ordered configurations, such as body-centered cubic and simple cubic packings, exhibit higher porosity of 0.32 and 0.48, respectively.

Yu et al. [5] investigated how mechanical vibration can be used to achieve the densest possible packing of spherical and non-spherical particles using DEM. Two primary methods are explored: in the first approach (Method I), particles are poured into a container and then subjected to vibration, while in the second approach (Method II), particles are added incrementally in small batches while vibration is applied simultaneously, as shown in Fig. 11.2. Using glass beads of 3 mm diameter, the experimental results demonstrate that traditional RCP could be surpassed with multi-dimensional vibration, reaching the density up to 0.661 for Method I. However, the most significant breakthrough comes with Method II, where optimized 3D vibration and small batch sizes enable the particles to self-assemble into FCC packing, achieving the theoretical maximum density of about 0.74. DEM simulations are conducted to provide deeper insights into the force structures and particle dynamics during vibration. The simulation results confirm that ordered force chains emerge in FCC packings, with each sphere maintaining 8–12 contacts depending on orientation.

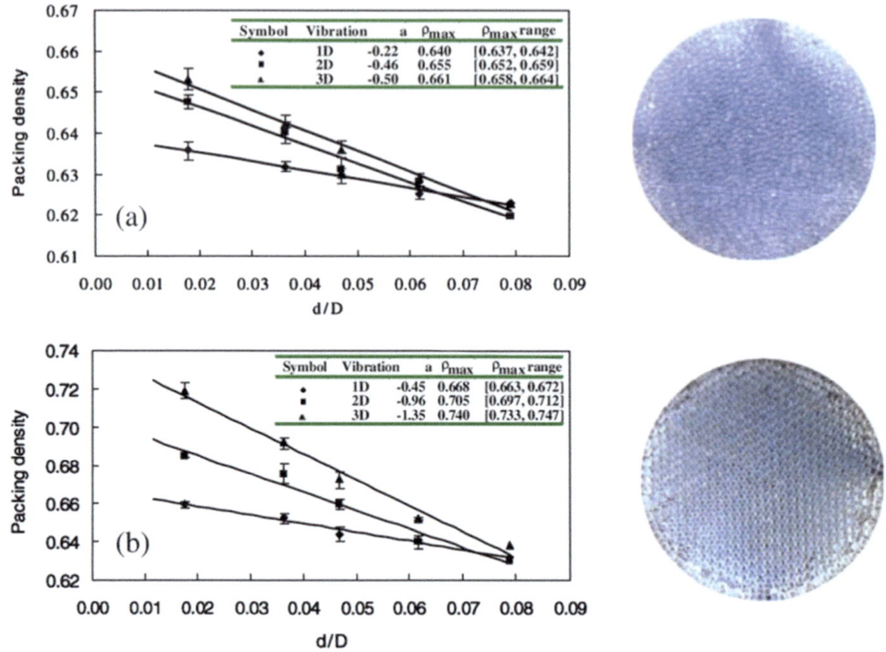

Fig. 11.2 Particle density as a function of d/D (ratio of particle diameter to cylindrical container) under different conditions: **a** method I; **b** method II [5]

While defects occasionally disrupt perfect crystallization, the method consistently produces densities far exceeding RCP.

Amirifar et al. [6] reported that the perfect ordered packing of mono-sized granular spheres can be obtained via random packing protocols for the first time. Figure 11.3 shows an inverted tetrahedron, which is a half-square-pyramid (HSP). Mono-sized spheres can spontaneously form a perfect single FCC crystal when randomly fed at a slow rate, achieving the theoretical maximum density of 0.74. This occurs because the container's geometry ensures that new particles settle only into stable lattice positions, aligning boundary and central structures. The DEM simulations also reveal the impact of feeding rate on defect formation. When particles are introduced in larger number of particles fed per batch N_B, rare point defects appear due to temporary arching, though defect rates remain below 0.1%. However, defect percentages increase sharply for $N_B > 20$, highlighting the importance of slow, controlled feeding for perfect crystallization. Additionally, simulations with varied material properties such as density and friction coefficients show that the self-assembly mechanism remains robust, producing FCC crystals regardless of particle type.

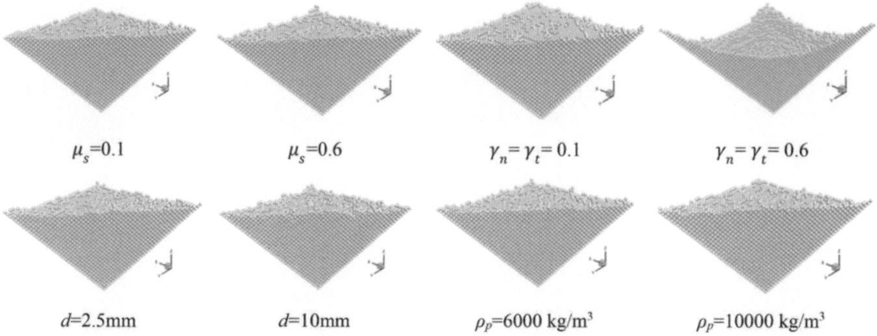

Fig. 11.3 Simulated self-assembly of mono-sized spheres with different material properties [6]

11.1.3 Non-spherical Particle Packing

Non-spherical particle packing introduces additional complexity due to geometric constraints and anisotropic interactions. DEM studies have examined various shapes, including ellipsoids, rods and platelets, which generally result in higher porosity compared to spherical particles. The packing of such particles often shows preferential alignment under shear or flow conditions, leading to direction-dependent properties. Another important aspect is polydisperse packing, where mixtures of different particle sizes can achieve higher densities as smaller particles fill the gaps between larger ones. DEM simulations of such systems are crucial for optimizing industrial processes involving powder mixtures, such as in concrete production or pharmaceutical manufacturing.

Fan et al. [7] investigated the role of particle-form polydispersity in the fabric of granular packings using DEM simulations with ellipsoidal particles. The research focuses on understanding how variations in particle shape such as elongation index (EI) and flatness index (FI) affect packing density, coordination number, contact forces and fabric anisotropy. An overall polydispersity index (OPI) is introduced to quantify the degree of particle-form variation, as shown in Fig. 11.4, and three series of particle-form distributions with varying median values, ranges and curve shapes are designed. The results reveal that in monodisperse packings, packing density and mean coordination number follow parabolic relationships with EI and FI, peaking at intermediate values before declining. Fabric anisotropy increases significantly as particles become more elongated or flattened, with FI having a more pronounced effect than EI. For polydisperse packings, higher OPI values lead to reduced packing density and mean coordination number, demonstrating that particle-form diversity substantially impacts packing behavior.

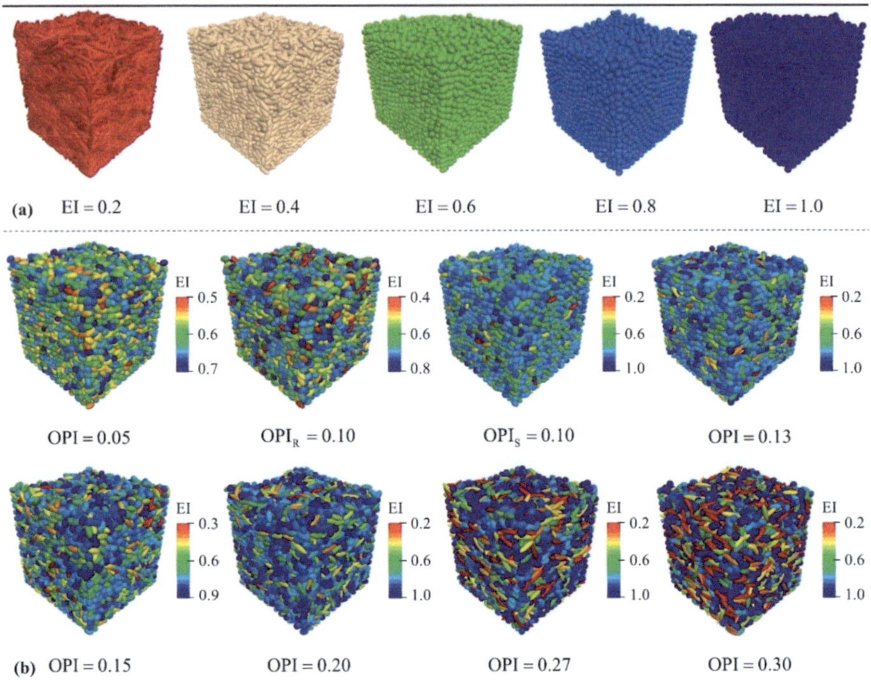

Fig. 11.4 Final obtained packings with various EI distributions: **a** monodisperse packings with different EI values and **b** polydisperse packings with different OPIs [7]

11.1.4 Dynamic and Process-Dependent Packing

Dynamic and process-dependent packing refers to the evolving arrangement of granular materials under external influences such as vibration. Unlike static packing, where particles settle into stable configurations, dynamic packing involves continuous rearrangement due to energy input. This process is highly sensitive to parameters like vibration amplitude, frequency, direction and particle shape.

Yu et al. [8] investigated the packing behaviors of trilobed particles under continuous vertical vibration using the DEM, as shown in Fig. 11.5. Trilobes, which are non-spherical particles commonly used as catalyst supports in chemical engineering, are modeled by the multisphere method. The research focuses on how vibration conditions including amplitude A and frequency f affect packing density, orientation distribution and the formation of ordered structures. It is found that moderate vibration conditions with $A = 1$–2 mm and $f = 20$–40 Hz promote dense packing. However, excessive vibration disrupts ordered structures, leading to looser packing. Results also show that under moderate conditions with $A = 2.0$ mm and $f = 20$ Hz, trilobes align predominantly parallel to the gravity direction, forming crystalline-like layers. In contrast, low-amplitude, high-frequency vibrations with $A = 1.0$ mm, $f =$

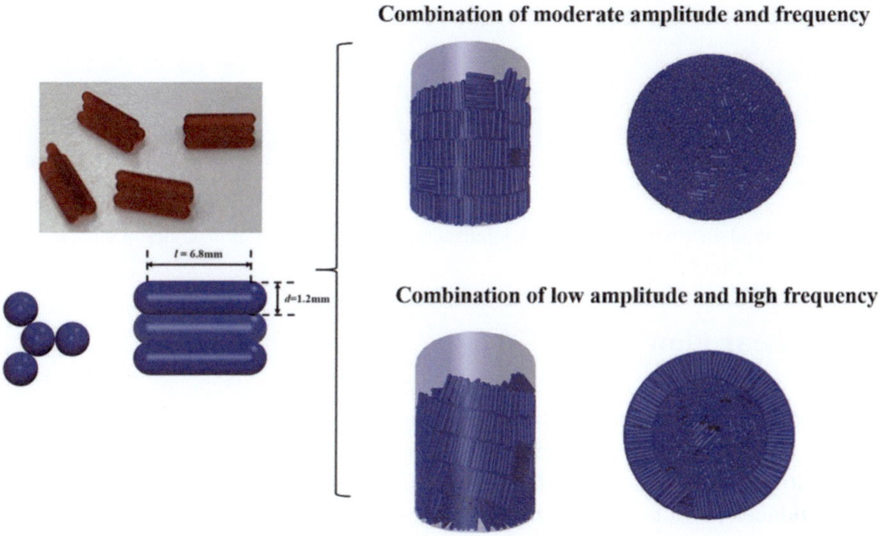

Fig. 11.5 Effect of vibration conditions on packing structures of trilobed particles [8]

40 Hz induce "sunflower-shaped" heterogeneous clusters, where central trilobes align vertically while surrounding ones orient horizontally. Despite these structural differences, both configurations exhibit similar overall packing densities, highlighting the need for additional metrics such as radial distribution functions, nematic order parameters to fully characterize packing behavior. The study also identifies two mechanisms driving ordered cluster formation: nucleation growth (small clusters forming near container walls) and engulfment growth (coalescence of clusters with identical orientations). Notably, interlocking effects between trilobes stabilize these clusters, but excessive vibration energy disrupts alignment, homogenizing the bed.

Yu et al. [9] also conducted DEM simulations to study the effect of vibration direction on the packing behavior of spherocylindrical particles. Key findings reveal that vibration direction significantly influences packing density and structure, as shown in Fig. 11.6. When vibrations are purely vertical, the highest packing density of 0.6835 is achieved, accompanied by the formation of ordered clusters. In contrast, other vibration directions result in lower densities and less ordered structures due to radial velocity fluctuations disrupting particle alignment. The study also highlights that contact modes between particles, particularly parallel edge-to-edge contacts, play a crucial role in densification.

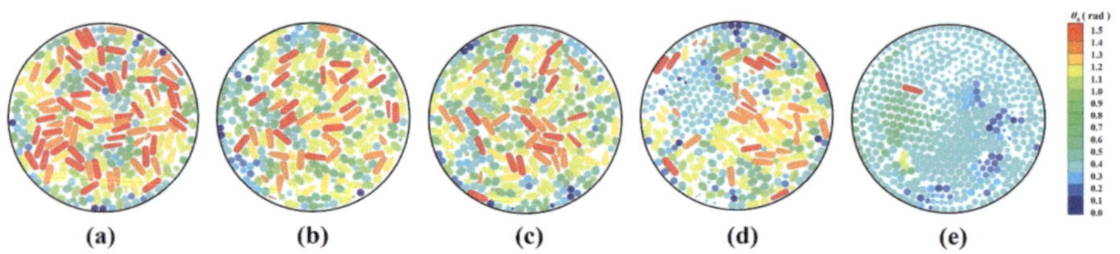

Fig. 11.6 Radial distributions of spherocylinders under different vibration angles: **a** 0.0°, **b** 22.5°, **c** 45.0°, **d** 67.5° and **e** 90.0°

11.2 Shear Flow

Shear flow refers to the motion of a granular material where adjacent layers move parallel to each other at different velocities, creating a velocity gradient perpendicular to the flow direction. This deformation arises due to applied shear stress, which induces particle reorganization, energy dissipation and structural transitions in some cases. In granular systems, shear flow can lead to phenomena like shear-induced ordering, where particles align to minimize resistance, or shear banding, where deformation localizes into narrow regions. The behavior depends on factors such as particle shape, friction and confinement.

Yang et al. [10] investigated the shear flow behavior of binary granular mixtures composed of large non-spherical particles and small spherical particles using the DEM, as shown in Fig. 11.7. The primary objective is to understand how particle shape, size distribution and solid volume fraction influence the macroscopic flow properties of granular materials. Results show that in dilute regimes, where particle collisions are infrequent, shear stresses decrease with increasing solid fraction due to reduced mean free paths between particles. However, in dense regimes, where particle collisions dominate, stresses increase with solid fraction as contact forces become more significant. The study also reveals that the relative proportion of large to small particles plays a crucial role. Higher fractions of large particles lead to increased shear stresses, as these particles contribute more significantly to momentum transfer. Interestingly, when the fraction of small particles is sufficiently high, the influence of large particles diminishes. Additionally, large non-spherical particles tend to orient more uniformly in the shear flow direction when small particles are abundant, thereby reducing shear stresses by minimizing perpendicular collisions. Particle shape and aspect ratio are found to significantly affect flow dynamics. Elongated particles with higher aspect ratios exhibit lower stresses in dilute flows due to shorter mean free paths but higher stresses in dense flows because of increased contact density. Additionally, oval-like particles, which have rougher surfaces, generate higher stresses than smoother rod-like particles in dense regimes due to stronger interparticle interactions.

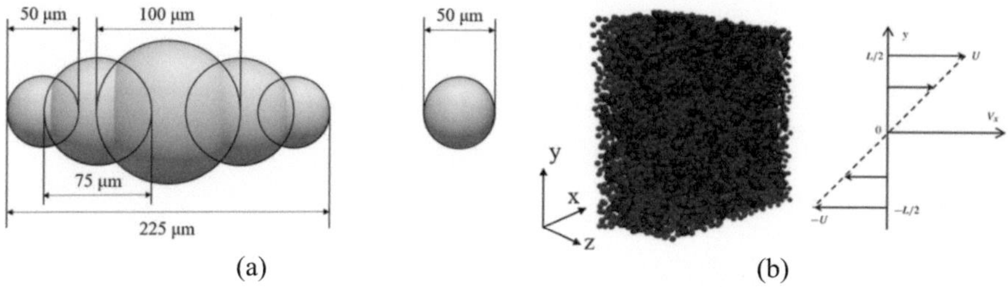

Fig. 11.7 **a** Binary mixture with large non-spherical particle and small spheres; **b** the numerical domain with binary mixtures [10]

11.3 Silo and Hopper Discharge

The discharge of bulk solids from silos and hoppers is a critical process in industries handling granular materials, where controlled and reliable flow is essential for operational efficiency. Silos, typically used for large-scale storage, rely on hoppers which have conical or wedge-shaped bottoms to facilitate gravity-driven discharge. The flow behavior depends on factors such as hopper geometry, material properties and external aids like vibrators or air injection systems. Two primary flow patterns dominate: mass flow, where all material moves uniformly, ensuring first-in-first-out discharge, and funnel flow, where a central channel forms, often leading to stagnant material and segregation. Challenges such as arching, ratholing and particle segregation can disrupt flow, necessitating design optimizations like steeper hopper angles, smoother surfaces or mechanical flow promoters. The applications of silo and hopper discharge can be seen in various sectors, including agriculture (grain storage), pharmaceuticals (powder dosing), construction (cement handling) and mining (ore transport). DEM simulations are revolutionizing silo and hopper design and operation, enabling predictive flow management and minimizing downtime.

Zhang et al. [11] investigated the flow mechanism transition in conical silos during the discharge of rice particles using DEM. The research focuses on identifying the critical height at which the flow pattern shifts from mass flow to funnel flow, as shown in Fig. 11.8. Key findings demonstrate significant changes in flow characteristics including velocity, porosity and particle orientation during this transition. Velocity shifts from uniform to uneven, porosity increases and particle orientation changes from horizontal to vertical as the flow transitions to funnel flow. The study also examines the impact of silo geometry parameters on the critical height. Results show that the critical height increases with larger hopper half angles or silo diameters due to greater flow resistance near the walls, while it decreases with larger outlet sizes because of reduced porosity and improved mass flow.

Najafi-Sani and Mansourpour [12] investigated polydisperse granular systems with wide particle size distributions in the silo discharge with a novel GPU-optimized solver, as shown in Fig. 11.9. Results show that for bidisperse silos, coarse particle discharge rates depend primarily on silo geometry, while fine particle discharge is dominated by percolation through coarse media. Flat-bottom silos exhibit slower

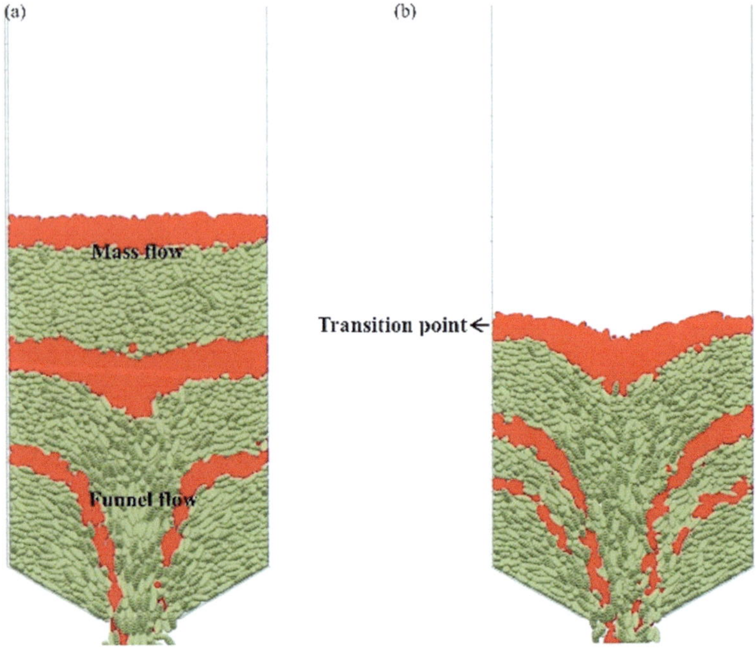

Fig. 11.8 Snapshots showing the transition process of the flow pattern: **a** initial discharge stage (mass flow at the top, funnel flow at the bottom) and **b** full funnel flow stage [11]

and less uniform discharge due to dead zones, whereas angled bottoms (30°–60°) enhance flow uniformity. As for the segregation effects, larger size ratios intensify fine particle percolation, leading to non-uniform outlet compositions. This effect is mitigated by steeper hopper angles, which delay segregation onset. Additionally, flat-bottom silos show significant residence time distribution deviations from ideal plug flow, while angled designs maintain more consistent flow patterns.

Dai et al. [13] investigated the evolution of force networks in a silo, focusing on mesoscale interparticle contact mechanics, as shown in Fig. 11.10. Key findings reveal that the force chain network is highly anisotropic, with force chain contacts constituting only 13–14% of total interparticle contacts, while strong contacts account for 37–41%. The collimation coefficient of force chains remains stable initially but decreases sharply over time, indicating reduced network stability. Both particle–particle and particle–wall friction coefficients significantly influence the number, strength and direction of force chains but have minimal impact on their length distribution. The results also show that force chains predominantly consist of 3–6 particles, with their strength decreasing as particles discharge from the reactor. The spatial distribution of force chains shows greater anisotropy compared to general contact forces, aligning more vertically due to gravitational and wall constraints. Friction coefficients play a critical role: higher interparticle friction increases force chain

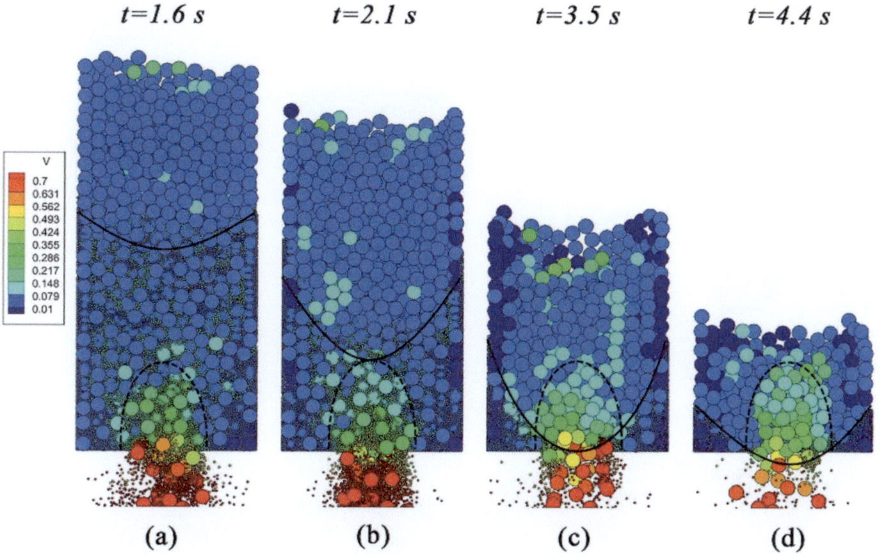

Fig. 11.9 Velocity color map of the discharge process for the flat-bottom silo with a size ratio of 7:1 [12]

stability and strength, while particle–wall friction primarily affects near-wall particle behavior.

Govender et al. [14] focused on the study of granular material flow in hoppers, emphasizing the impact of particle shape non-uniformity and polydispersity using the Blaze-DEM GPU code. Key findings include the significant influence of particle shape and size distribution on hopper discharge rates, with polyhedral particles generally discharging slower than spherical ones due to higher flow resistance and angularity, as shown in Fig. 11.11. The research also highlights the formation of stable boundary arches in low-angle hoppers, which reduce discharge rates by narrowing

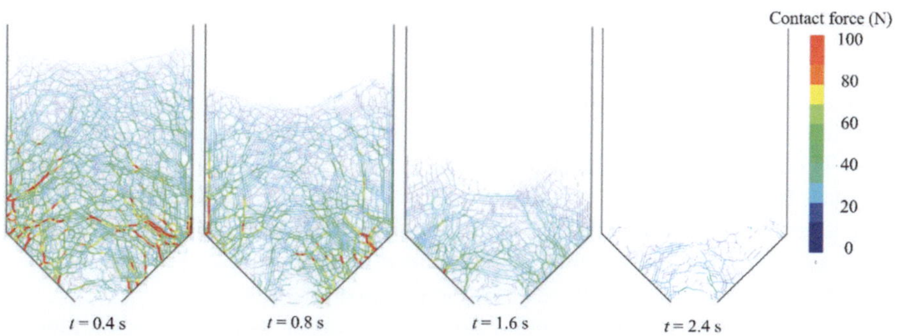

Fig. 11.10 Distributions of interparticle contact forces over time [13]

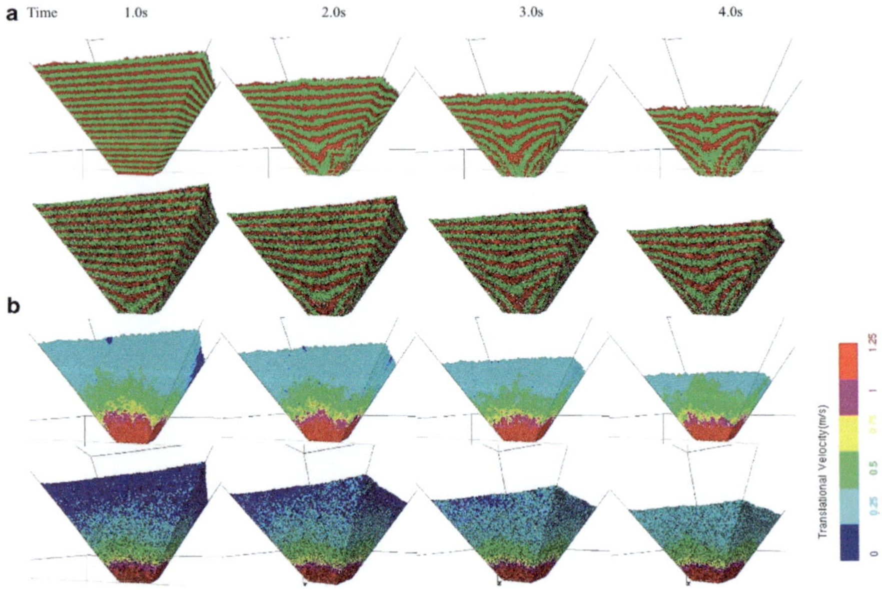

Fig. 11.11 Spatial layer profile discharge rate for the uniform shape mono-dispersed **a** spherical and polydispersed truncated tetrahedra particle systems and corresponding, **b** translational velocity magnitude profiles for a 60° hopper configuration [14]

the effective throat area. Additionally, the study underscores the computational efficiency of GPU-based DEM simulations, enabling large-scale analyses with up to one million polyhedral particles, a significant advancement over previous limitations of 10,000 particles.

11.4 Moving Beds

Moving beds are a type of dense granular system where solid particles flow continuously under gravity or mechanical forces while interacting with gases or liquids. Unlike fluidized beds that suspend particles through fluid flow, moving beds maintain a packed or semi-packed structure, allowing for controlled and steady-state operations. This design is particularly advantageous for processes requiring efficient heat and mass transfer, chemical reactions or material transport. The continuous flow of particles ensures consistent processing, while the gravity-driven mechanism reduces energy consumption, making moving beds both efficient and scalable for applications ranging from laboratory experiments to large industrial systems.

One of the most prominent applications of moving beds is in chemical looping combustion (CLC), a technology used for carbon capture with minimal energy loss. In CLC systems, moving beds function as air reactors [15–17] or fuel reactors [18], where oxygen carriers, such as metal oxides, circulate between reactors to transfer oxygen for combustion without direct air–fuel mixing. This process significantly reduces greenhouse gas emissions and improves energy efficiency. Moving beds are also widely utilized in catalytic processes, including fluid catalytic cracking and catalytic reforming, where catalyst particles flow continuously through reaction zones, enabling efficient regeneration and prolonged catalyst life. In addition to combustion and catalysis, moving beds play a critical role in adsorption and separation processes. They are employed in pressure swing adsorption systems and gas purification technologies, such as CO_2 capture, where adsorbent materials like zeolites move through columns to selectively separate gases. The thermal treatment industry also benefits from moving beds, particularly in rotary kilns and pyrolyzers used for processing biomass, ores and waste materials. These systems ensure uniform heating and precise reaction control, making them ideal for applications like waste-to-energy conversion and metal ore reduction. The mining and metallurgy sectors leverage moving beds in blast furnaces and direct reduction reactors, where descending ore particles react with ascending gases to produce metals like iron. This method enhances reaction efficiency and reduces energy consumption compared to traditional smelting processes. Furthermore, moving beds are integral to pharmaceutical and food processing, where they facilitate controlled drying, coating and granulation of particles, ensuring consistent product quality in pellet or tablet production.

Despite their advantages, moving beds present challenges such as particle attrition, flow instability and the need for precise control over friction and particle properties. Issues like bridging or arching can disrupt flow, requiring careful design and operational adjustments. Several typical application examples of DEM in moving beds are described below.

11.4.1 Chemical Looping Combustion

Shao et al. [19] presented a comprehensive study on the flow characteristics and solids' residence time distribution in a moving bed air reactor designed for CLC using a quasi-2D CFD-DEM cold flow model. Key findings reveal that increasing the loop seal gas velocity enhances the solids' mass flux while reducing the average residence time. The reactor angle significantly influences particle movements near the wall, whereas the wedge angle primarily affects particles in the lower reactor region. Smaller reactor angles, larger wedge angles and increased down-comer diameters promote flow patterns closer to ideal plug flow. The study also highlights the role of internals in accelerating near-wall particles, with triangular arrays proving more effective than square arrays in slowing central particles and improving flow uniformity, as shown in Fig. 11.12. Additionally, the mass flow index is used to distinguish between mass flow and funnel flow regimes, showing that geometric adjustments can optimize flow patterns.

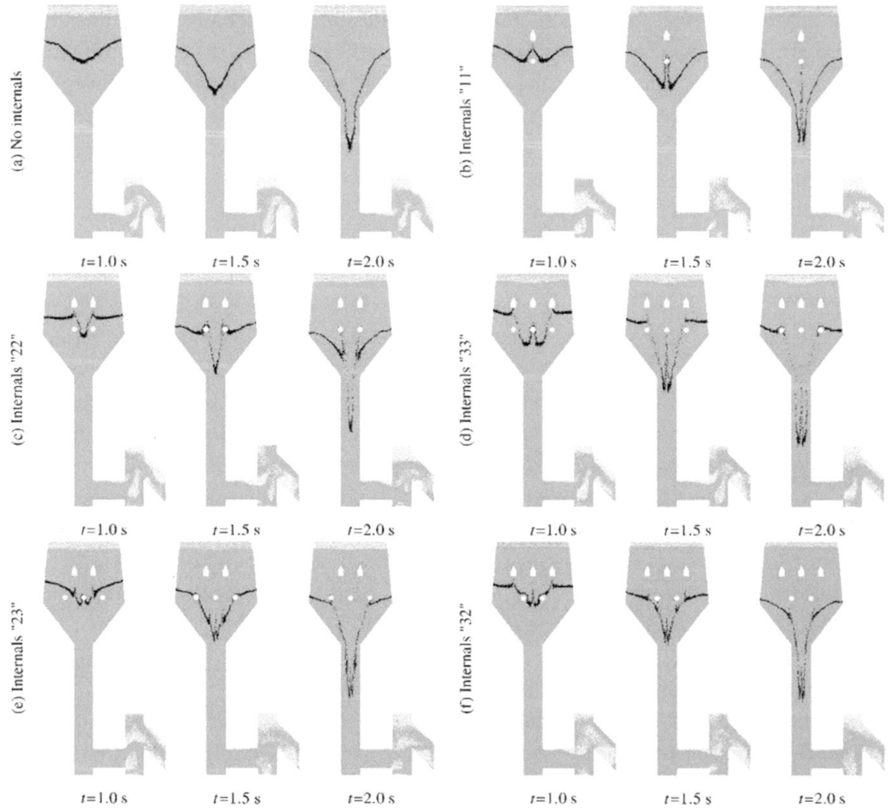

Fig. 11.12 Instantaneous tracer particle distributions in the moving bed air reactor without internals and reactors with internals [19]

11.4.2 Pyrolysis

Zhang et al. [20] investigated a moving bed pyrolysis reactor designed for the clean and resourceful treatment of oil sludge, a hazardous waste rich in hydrocarbons. The reactor features a rectangular upper section and a conical lower section, measuring 1.1 m in height and 0.35 m in length and width. It includes three layers of angular tubes internally, strategically positioned to ensure uniform particle descent and carrier gas distribution. The reactor employs solid heat carriers (oil sludge ash) to enhance heat transfer efficiency and nitrogen as the carrier gas to minimize secondary cracking of pyrolysis products. The design aims to optimize the flow and mixing uniformity of oil sludge particles (OS) and heat carriers (HC), which are critical for efficient pyrolysis. The research focuses on analyzing the gas–solid–solid three-phase flow (carrier gas, HC and OS) in the moving bed reactor using the CFD-DEM approach. The Euler–Euler–Lagrange multiphase flow model is employed, where the carrier gas and HC are treated as continuous phase, while OS particles are tracked individually. This hybrid model captures interactions between phases, including drag forces, particle collisions and momentum transfer, to simulate the complex dynamics within the reactor. The simulation results identify optimal operating conditions for the moving bed reactor: a 1:1 OS:HC mass ratio and 7.5 L/min carrier gas flow rate for optimal flow uniformity, and a 1:4 OS:HC ratio with 5 L/min gas flow for best mixing, as shown in Fig. 11.13. These conditions ensure efficient pyrolysis by minimizing particle segregation and maximizing heat transfer.

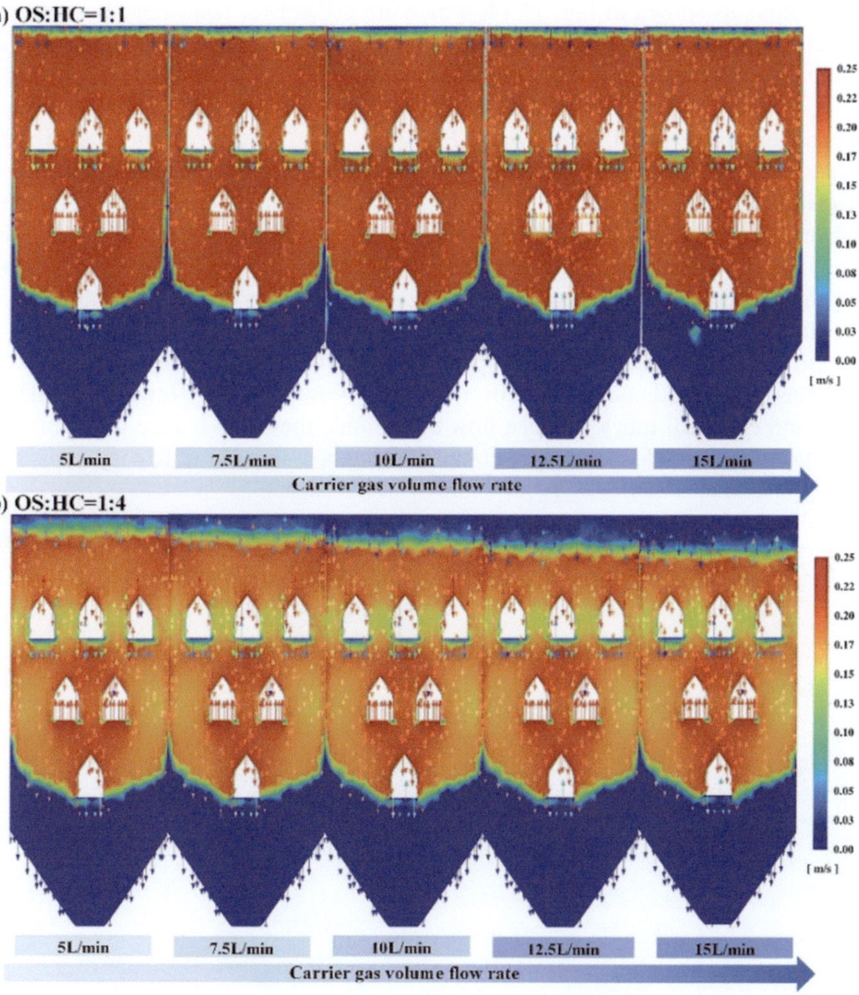

Fig. 11.13 Effect of carrier gas volume flow rate on carrier gas velocity distribution in a moving bed pyrolysis reactor: **a** OS:HC = 1:1; **b** OS:HC = 1:4 [20]

11.4.3 Heat Exchanger

The application of moving bed heat exchangers is primarily focused on efficiently recovering waste heat from high-temperature granular materials, such as industrial by-products or solar thermal storage media.

Bartsch and Zunft [21] presented a comprehensive investigation into the granular flow dynamics around horizontal tubes in moving bed heat exchangers using DEM. The simulations predict the formation of a stagnant zone above the tubes and a void area below, as shown in Fig. 11.14, which are critical for heat transfer efficiency. Additionally, wall friction reduces flow velocities near the transparent front wall, while the bulk flow exhibits higher velocities and more pronounced void effects.

Li et al. [22] presented an optimization study for tubular moving bed heat exchangers using DEM, as shown in Fig. 11.15. The study focuses on addressing issues such as stagnant zones, cavity zones and bridging-and-breaking phenomena around heat exchange tubes, which significantly hinder heat transfer efficiency between high-temperature solid granules and tube walls. Three types of tubes including cylindrical, elliptical and a newly designed gradually shrinking and expanding (GSE) tube are compared to evaluate their impact on flow stability and heat transfer performance. Key findings reveal that the GSE tube design, inspired by the circular orifice of cylindrical hoppers, reduces vertical-direction forces between particles and tube bundles, thereby improving flow stability. However, while the GSE and elliptical tubes enhance flow conditions, they also reduce the number of particle-tube contacts, potentially compromising heat transfer efficiency.

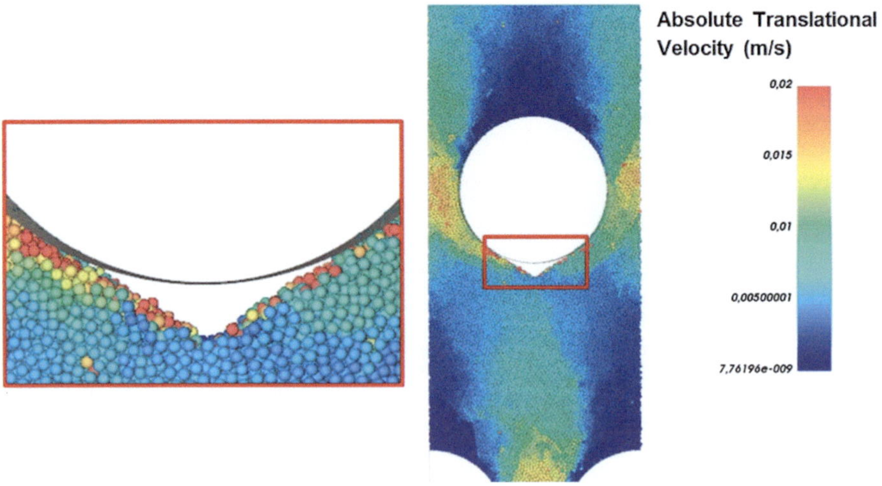

Fig. 11.14 Example particle assembly from DEM simulation of a particle heat exchanger. Particle color shown according to their translational velocity. Left: Void area below the tube [21]

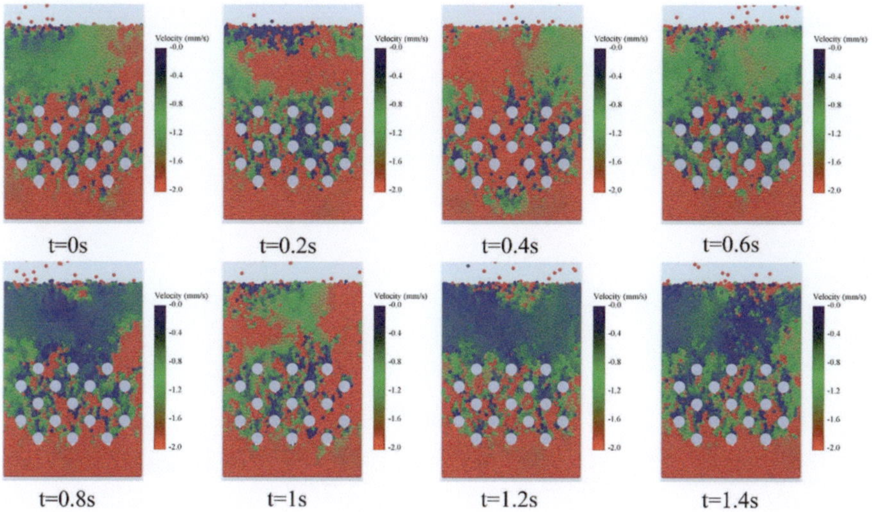

Fig. 11.15 Particle flow velocity distribution at the tube bundle area in the cylindrical tube moving bed heat exchanger [22]

11.4.4 Blast Furnace

A blast furnace is a large, vertical, counter-current chemical reactor and heat exchanger used primarily for the smelting of industrial metals from their ores, most commonly iron. DEM is a useful tool to simulate the behavior of bulk materials like ore and coke composed of millions of individual particles in the reactor.

Yang et al. [23] investigated heat transfer and gas–solid flow in a moving bed reactor using CFD-DEM approach, specifically focusing on conditions relevant to the raceway region of an ironmaking blast furnace, as shown in Fig. 11.16. Key findings reveal that particle–fluid convection dominates heat transfer, contributing approximately 80% of the total heat flux, while particle–particle conduction accounts for about 20%, and radiation plays a negligible role at lower temperatures lower than 51 °C. However, at high temperatures over 2000 °C, radiation becomes significant, contributing up to 12% of the heat flux. The study also identifies distinct flow regions in the moving bed: a fast-flow zone near the particle discharge outlet, a stagnant zone with minimal particle movement and a transition zone where particles mix. The stagnant zone's size is influenced by particle size, with smaller particles leading to a larger stagnant region. Gas flow rate significantly affects temperature distribution but has minimal impact on solid flow patterns. Conversely, increasing the solid flow rate reduces the stagnant zone size and lowers the bed temperature due to shorter particle residence times.

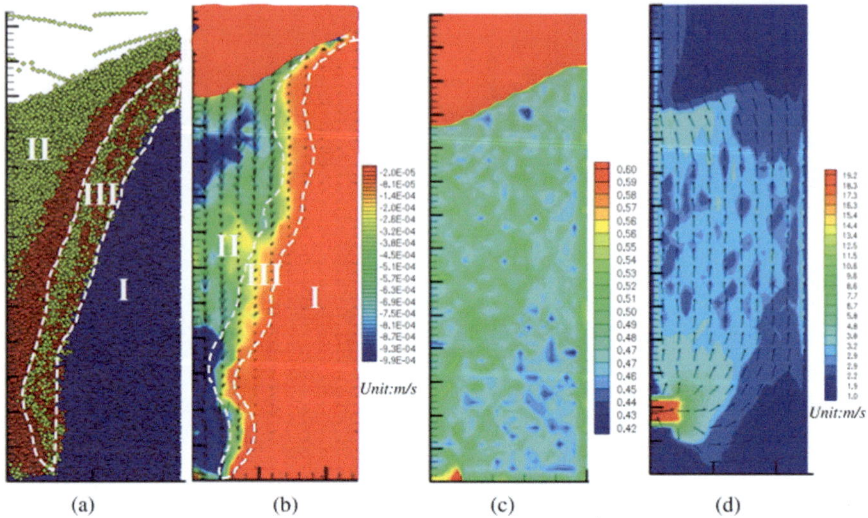

Fig. 11.16 Solid flow pattern with gas (**a**), solid velocity (**b**), the spatial distribution of porosity (**c**) and the gas velocity field (**d**) when solid flow rate is 0.086 kg/s, gas flow rate is 0.08 kg/s and time $t = 112$ s [23]. Note that three regions are corresponding to stagnant zone(I), fast-flow zone(II) and transit zone(III)

References

1. L.E. Silbert, Jamming of frictional spheres and random loose packing. Soft Matter **6**(13), 2918–2924 (2010). https://doi.org/10.1039/c001973a
2. L.T. To, Z.H. Stachurski, Random close packing of spheres in a round cell. J. Non-Cryst. Solids **333**(2), 161–171 (2004). https://doi.org/10.1016/j.jnoncrysol.2003.09.041
3. K. Washino, E.L. Chan, D. Faroux, T. Tsuji, T. Takahashi, S. Sasabe, On DEM simulation of loose packing behaviour of fine and cohesive particles. Adv. Powder Technol. **36**(3), 4809 (2025). https://doi.org/10.1016/j.apt.2025.104809
4. X. An, R. Yang, K. Dong, A. Yu, DEM study of crystallization of monosized spheres under mechanical vibrations. Comput. Phys. Commun. **182**(9), 1989–1994 (2011). https://doi.org/10.1016/j.cpc.2011.02.006
5. A.B. Yu, X.Z. An, R.P. Zou, R.Y. Yang, K. Kendall, Self-assembly of particles for densest packing by mechanical vibration. Phys. Rev. Lett. **97**(26), 5501 (2006). https://doi.org/10.1103/PhysRevLett.97.265501
6. R. Amirifar, K. Dong, A. Yu, Ordered packing of uniform spheres via random packing protocol. Powder Technol. **409**, 7853 (2022). https://doi.org/10.1016/j.powtec.2022.117853
7. M. Fan, D. Su, Y. Li, X. Chen, Exploring the role of particle-form polydispersity in the fabric of granular packing: insights from DEM simulations with ellipsoidal particle assemblies. Powder Technol. **435**, 9402 (2024). https://doi.org/10.1016/j.powtec.2024.119402
8. F. Yu, D. Cui, Y. Zheng, G. Zhou, Effect of vibration conditions on packing structures of trilobed particles. Powder Technol. **438**, 9643 (2024). https://doi.org/10.1016/j.powtec.2024.119643
9. F. Yu, D. Cui, Y. Zhang, G. Zhou, Effect of vibration direction on the packing of spherocylinders. Adv. Powder Technol. **34**(9), 4138 (2023). https://doi.org/10.1016/j.apt.2023.104138

10. J. Yang, Y. Guo, K.E. Buettner, J.S. Curtis, DEM investigation of shear flows of binary mixtures of non-spherical particles. Chem. Eng. Sci. **202**, 383–391 (2019). https://doi.org/10.1016/j.ces.2019.03.027
11. Y. Zhang, F. Jia, Y. Zeng, Y. Han, Y. Xiao, DEM study in the critical height of flow mechanism transition in a conical silo. Powder Technol. **331**, 98–106 (2018). https://doi.org/10.1016/j.powtec.2018.03.024
12. M.A. Najafi-Sani, Z. Mansourpour, A new GPU-based DEM simulator for polydispersed granular systems with wide size distribution and its application in the silo discharge. Adv. Powder Technol. **33**(8), 3704 (2022). https://doi.org/10.1016/j.apt.2022.103704
13. W. Dai, Y. Shao, S. Yin, T. Song, R.K. Agarwal, Numerical investigation of force network evolution in a moving bed air reactor. Fire **7**(11), 376 (2024). https://doi.org/10.3390/fire7110376
14. N. Govender, D.N. Wilke, P. Pizette, N.E. Abriak, A study of shape non-uniformity and polydispersity in hopper discharge of spherical and polyhedral particle systems using the Blaze-DEM GPU code. Appl. Math. Comput. **319**, 318–336 (2018). https://doi.org/10.1016/j.amc.2017.03.037
15. Y. Shao, X. Wang, B. Jin, Z. Kong, X. Wang, Y. Zhang, Gas–solid hydrodynamics of an iG-CLC system with a two-stage counter-flow moving bed air reactor. Chem. Eng. Res. Des. **143**, 100–113 (2019). https://doi.org/10.1016/j.cherd.2019.01.004
16. Y. Shao, R.K. Agarwal, X. Wang, B. Jin, Numerical simulation of a 3D full loop iG-CLC system including a two-stage counter-flow moving bed air reactor. Chem. Eng. Sci. **217**, 115502 (2020). https://doi.org/10.1016/j.ces.2020.115502
17. Y. Shao, Y. Hu, X. Wang, B. Jin, R.K. Agarwal, Numerical study of a chemical looping combustion system: effects of fuel type and particle size distributions on the operating performance. Chem. Eng. J. **500**, 476 (2024). https://doi.org/10.1016/j.cej.2024.156476
18. D. Wang et al., Coal direct chemical looping combustion process: Design and operation of a 25-kWth sub-pilot unit. Fuel **108**, 370–384 (2012). https://doi.org/10.1016/j.fuel.2012.12.038
19. Y. Shao, R.K. Agarwal, J. Li, X. Wang, B. Jin, Computational fluid dynamics-discrete element model simulation of flow characteristics and solids' residence time distribution in a moving bed air reactor for chemical looping combustion. Ind. Eng. Chem. Res. **59**(40), 18180–18192 (2020). https://doi.org/10.1021/acs.iecr.0c02426
20. T. Zhang, D. Jiao, X. Wang, L. Xia, CFD-DEM analysis of gas-solid-solid three-phase flow in a moving bed pyrolysis reactor. Powder Technol. **458**, 65–75 (2025). https://doi.org/10.1016/j.powtec.2025.120948
21. P. Bartsch, S. Zunft, Granular flow around the horizontal tubes of a particle heat exchanger: DEM-simulation and experimental validation. Sol. Energy **182**, 48–56 (2019). https://doi.org/10.1016/j.solener.2019.01.086
22. Z. Li, W. Lu, T. Zhou, Y. Liu, Y. Zhang, An optimization tube design for the tubular moving bed heat exchanger (MBHE) by the analogy with hopper discharge process. Case Stud. Therm. Eng. **71**, 321–543 (2025). https://doi.org/10.1016/j.csite.2025.106193
23. W.J. Yang, Z.Y. Zhou, A.B. Yu, Particle scale studies of heat transfer in a moving bed. Powder Technol. **281**, 99–111 (2015). https://doi.org/10.1016/j.powtec.2015.04.071

Chapter 12
Applications to Aerodynamic Systems

12.1 Vertical Pneumatic Conveying System

Vertical pneumatic conveying systems are designed to transport bulk granular materials or powders in a vertical direction using a gas flow, typically air. These systems operate by utilizing aerodynamic drag forces and pressure differentials to lift or lower particles through pipelines, overcoming gravitational effects. The design and operation of such systems vary significantly depending on whether they function in dilute-phase mode, characterized by high gas velocity and low solid concentration, or dense-phase mode, which features higher solid loading at lower gas velocities.

The application of vertical pneumatic conveying can be seen in numerous industries where efficient vertical transport is essential due to spatial constraints or process requirements. In the chemical and pharmaceutical sectors, these systems are crucial for handling powders in reactors, dryers and packaging units. The food industry relies on them for moving ingredients like flour, grains and powdered additives in processing and storage facilities. Mining and mineral processing operations use vertical pneumatic systems to elevate materials such as ores, cement and fly ash. Energy and environmental applications include biomass circulation in combustion systems and the transport of particulate matter in carbon capture processes.

12.1.1 Research Focus

Research on vertical pneumatic conveying systems has gained significant attention due to the complex interplay of gas–solid dynamics in vertical transport applications. The primary focus of current investigations centers on understanding and optimizing the fundamental mechanisms governing particle movement, system efficiency and operational reliability. A major area of study involves characterizing flow regimes and

R. K. Agarwal and Y. Shao, *Theory and Application of Discrete Element Method to Simulation of Multiphase Flows*, https://doi.org/10.1007/978-3-032-11157-9_12

their transitions, particularly the delicate balance between dilute-phase and dense-phase flows. Researchers employ advanced experimental techniques and numerical simulations to identify critical velocity thresholds that prevent particle settling while minimizing energy consumption. The hydrodynamics of particle-gas interactions in vertical transport presents another critical research frontier. Scientists are particularly interested in how particle size distribution, shape and density affect suspension and conveyance characteristics. Energy efficiency and pressure drop minimization constitute a persistent research challenge in vertical pneumatic conveying. Investigations focus on quantifying and reducing energy losses associated with particle–wall friction, air compression and gravitational resistance. Additionally, material degradation and equipment wear present significant practical concerns that drive substantial research efforts. Studies examine the abrasive effects of different particle types on pipeline materials, with particular attention to bend sections and other high-wear areas.

12.1.2 Example Applications

Lim et al. [1] conducted a comprehensive numerical investigation of pneumatic conveying of granular materials in both vertical and horizontal pipes using CFD-DEM approach. The research focuses on simulating and analyzing different flow regimes that emerge under varying operating conditions, including gas velocities and solid concentrations. In vertical pneumatic conveying, two primary flow regimes are identified: dispersed flow and plug flow, as shown in Fig. 12.1. The dispersed flow regime, characterized by particles distributed uniformly throughout the pipe, is observed at high gas velocities and low solid concentrations. Conversely, the plug flow regime, where particles cluster into dense plugs, dominates at low gas velocities and high solid concentrations. The solid concentration profiles for these regimes reveal a symmetrical but non-uniform distribution for dispersed flow and a flat distribution for plug flow. Hysteresis in solid flow rates is noted near the transition between these regimes, indicating path-dependent behavior. For horizontal pneumatic conveying which will be introduced in the following section, additional flow regimes emerge due to gravitational effects, as shown in Fig. 12.2. At low gas velocities and solid concentrations, stratified flow and moving dunes regimes are observed, where particles settle near the bottom wall and form dynamic clusters, respectively. Higher gas velocities lead to homogeneous flow, while elevated solid concentrations result in slug flow, resembling plug flow in vertical pipes. The solid concentration profiles for stratified flow show higher concentrations near the bottom wall, while slug flow exhibits a flat profile similar to plug flow.

To understand the slug flow in vertical pneumatic conveying systems, Wang et al. [2] employed CFD-DEM approach to analyze the formation, evolution and stability of slugs under varying operational conditions, as shown in Fig. 12.3. The simulations reveal that slugs form when particles accumulate and are entrained upward by gas flow, with particles at the slug base experiencing lower stress and eventually

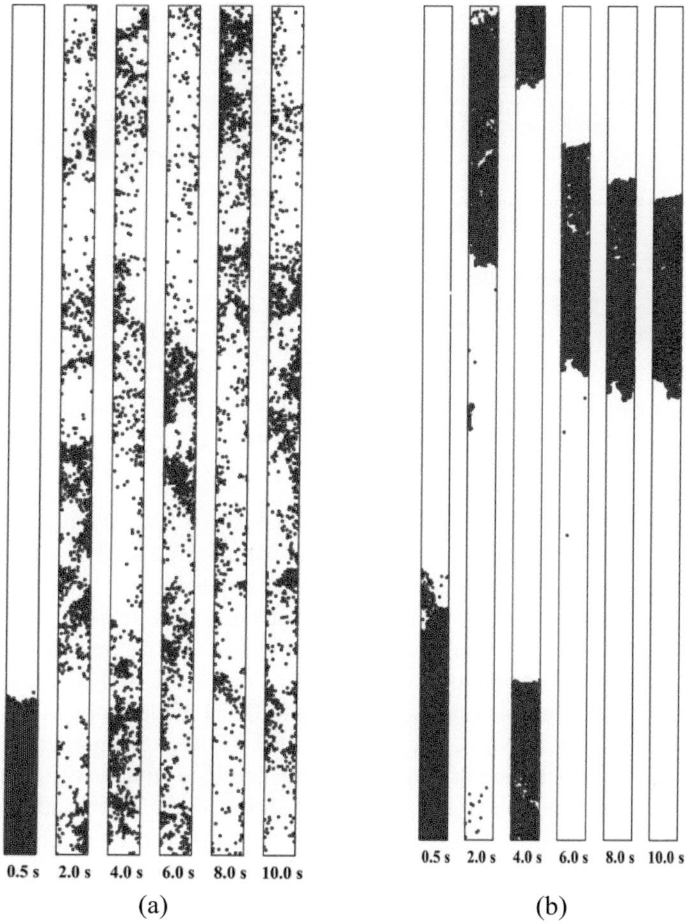

Fig. 12.1 Vertical pneumatic conveying in two flow regimes: **a** dispersed flow; **b** plug flow regime [1]

separating to merge with subsequent slugs. Gas velocities have great influence on the slug dynamics, with higher velocities reducing slug volume but increasing slug velocity and frequency. Larger riser diameters are found to increase slug volume but decrease velocity due to diminished gas-particle momentum exchange. Particle size also plays a significant role, with larger particles forming more stable slugs at lower velocities, while smaller particles exhibited higher velocities but greater instability. Force analysis in the slug flow reveals that drag and pressure gradient forces peak in the middle and upper sections of the riser, driving slug motion. The study also examines flow field characteristics, showing that slug passage induces significant fluctuations in solid concentration, gas velocity and pressure. Gas velocity spikes

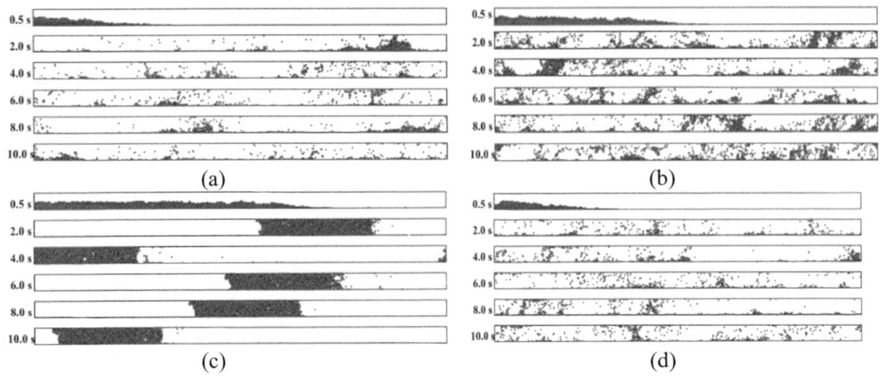

Fig. 12.2 Horizontal pneumatic conveying in four flow regimes: **a** stratified flow regime; **b** dune flow regime; **c** slug flow regime; **d** homogeneous flow regime [1]

occur as slugs constrict the flow channel, analogous to a nozzle effect, while pressure fluctuations exhibit multipeak spectra in power spectral analysis, reflecting the dynamic processes of slug formation, merger and dispersion.

Yang et al. [3] investigated the suspension behavior of large non-spherical particles (5–30 mm) in vertical pneumatic conveying systems, as shown in Fig. 12.4. The research focuses on understanding how particle shape and rotation influence suspension dynamics, employing a coupled CFD-DEM methodology to analyze the complex gas–solid interactions. The non-spherical particles examined exhibit irregular geometries where the center of mass does not coincide with the geometric center, leading to rotational motion when subjected to fluid forces. This rotation generates local vortices and turbulence that significantly affect the suspension characteristics. The theoretical framework divides the suspension behavior into three distinct flow regimes based on Reynolds number: viscous frictional ($Re \leq 5.76$), transition ($5.76 < Re \leq 516.5$) and pressure differential ($Re > 516.5$) regimes. A key finding reveals that particle rotation in non-spherical particles reduces the required suspension velocity compared to spherical particles, as the dynamic variation in windward area alters the drag characteristics. Additionally, large particles (20–30 mm) create significant flow disturbances, forming localized high-pressure zones that modify the suspension requirements.

12.2 Horizontal Pneumatic Conveying System

Unlike vertical pneumatic conveying systems that must overcome gravity, horizontal configurations move materials along a predominantly level plane, though they often incorporate slight inclines or declines to facilitate material flow. These systems operate on the principle of maintaining sufficient air velocity to keep particles suspended in the airstream while preventing settling or pipeline blockage. The

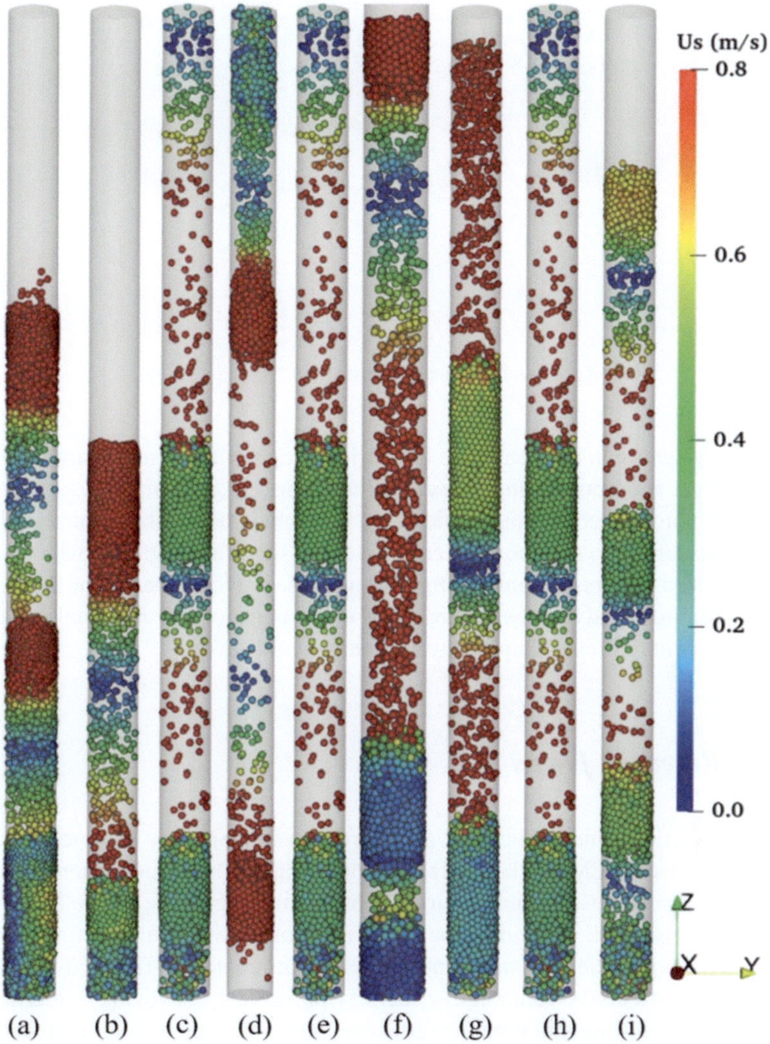

Fig. 12.3 Slug characteristics under different operating conditions: **a** particle diameter $d_p = 4$ mm; **b** particle diameter $d_p = 5$ mm; **c** particle diameter $d_p = 6$ mm; **d** rise diameter $Dr = 45$ mm; **e** rise diameter $Dr = 50$ mm; **f** rise diameter $Dr = 55$ mm; **g** superficial gas velocity $U_g = 7.39$ m/s; **h** superficial gas velocity $U_g = 8.87$ m/s; **i** superficial gas velocity $U_g = 10.15$ m/s [2]

design typically includes a material feed point, horizontal piping network, air mover such as a fan or compressor and a separation device at the discharge end. Two primary modes of operation exist: dilute-phase systems that use high air velocities with low solid-to-air ratios and dense-phase systems that employ lower velocities with higher material concentrations, often using pulsed airflow or other specialized techniques to move material slugs through the pipeline. The applications of horizontal pneumatic

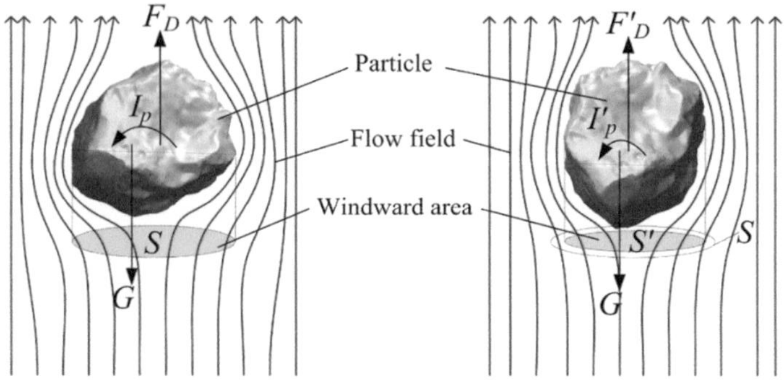

Fig. 12.4 Particle rotation process in vertical flow field [3]

conveying systems can be seen in the industries of agricultural processing, chemical manufacturing, food processing, mineral processing, etc. Additionally, what makes horizontal pneumatic conveying particularly valuable is its flexibility in plant layout design and its ability to transport materials over considerable distances which is up to several hundred meters in some configurations. The enclosed nature of the systems provides significant advantages in terms of dust control, environmental protection and product containment compared to mechanical conveyors.

12.2.1 Research Focus

Similar as vertical pneumatic conveying systems, a primary research focus of the application of DEM in the horizontal pneumatic conveying systems involves understanding particle–fluid interactions in horizontal pipelines, where DEM simulations precisely track individual particle trajectories while accounting for drag forces, particle–particle collisions and particle–wall contacts. These investigations have proven particularly valuable for identifying critical conveying velocities that prevent particle settling and pipeline blockage, with DEM providing unprecedented insight into how particle size distribution, shape and surface properties influence the minimum transport conditions. Additionally, a significant portion of DEM research focuses on dense-phase horizontal conveying, where particle–particle interactions dominate the flow physics. The formation and propagation of particle slugs are studied, examining how interparticle friction and cohesion affect slug stability and motion. These simulations reveal intricate details about stress distributions in particle assemblies during transport, helping engineers design systems that minimize energy consumption while maintaining reliable material flow.

12.2.2 Example Applications

Watano et al. [4] employed a two-dimensional DEM to analyze particle dynamics during horizontal pneumatic conveying and developed a simplified model to predict electrostatic charging based on collision mechanics. The study focuses on understanding how particle–wall collisions contribute to charge accumulation, with implications for industrial safety and process optimization. The proposed electrification model hinges on two key factors: the vertical component of particle collision velocity against the pipe wall and the number of such collisions. The results indicate that higher air velocities initially lead to increased charge accumulation due to more energetic collisions, but beyond a certain velocity, the charge decreases as particles spend less time in contact with the wall. The study also explores the effect of varying the initial particle load on electrification. Simulations with different numbers of particles demonstrate that higher particle loads reduce the specific charge per particle, as the increased frequency of particle–particle collisions shield the pipe wall from direct impacts.

Erken et al. [5] carried out coarse-grained CFD-DEM modeling for pneumatic conveying, particularly horizontal plug flow, focusing on reducing computational costs while preserving key physical behaviors. Various scaling methods for contact and drag forces are evaluated, aiming to develop a reliable methodology for industrial applications. Key findings highlight that cubic drag force scaling yields results closest to the original simulations, whereas linear and quadratic scaling lead to unrealistic air velocities. For contact forces, the constant relative overlap method preserves contact stresses effectively and allows larger DEM time steps, enhancing computational efficiency. In contrast, scaling the sliding friction coefficient results in poor accuracy. The study also examines the impact of coarse-graining (CG) ratios on plug properties, such as porosity, velocity and particle exchange rates, as shown in Fig. 12.5. Higher coarse-graining ratios increase plug porosity and reduce spatial resolution, affecting plug dynamics. The constant relative overlap model outperforms other methods in maintaining contact pressures, especially at higher coarse-graining ratios. Additionally, smoothing exchange fields improves stability when particle sizes approach CFD grid dimensions, though it may influence results at extreme coarse-graining levels.

12.3 Inclined Pneumatic Conveying System

Inclined pneumatic conveying systems bridge the operational characteristics of both horizontal and vertical systems while presenting their own distinct challenges. These systems transport granular or powdered materials through pipelines set at angles typically ranging from 15° to 70° from horizontal, using air or other gases

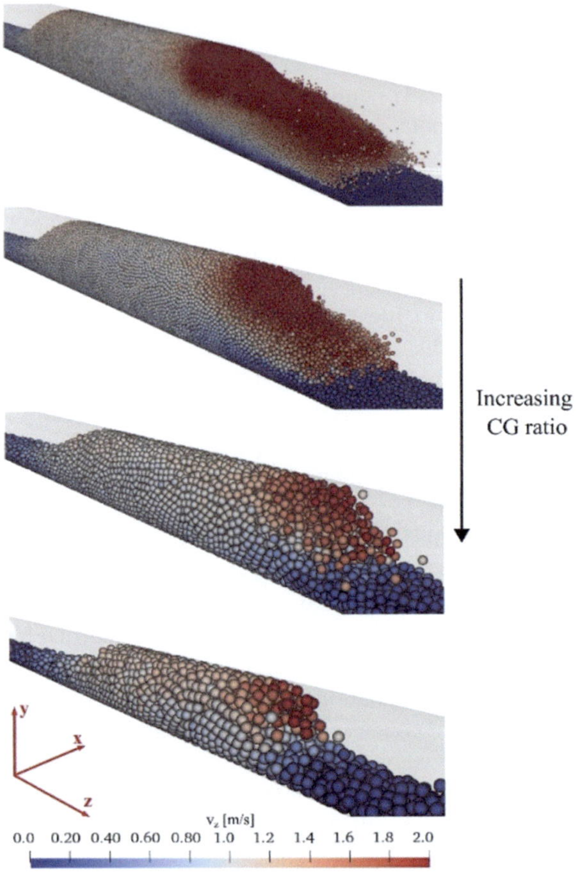

Fig. 12.5 Snapshots taken at different coarse-graining ratios in horizontal plug flow pneumatic conveying [5]

as the conveying medium. The fundamental principle involves maintaining sufficient air velocity to overcome both gravitational forces and particle–wall friction while preventing material backflow or pipeline blockage. Unlike purely horizontal conveying where gravity plays a minimal role, or vertical systems where it acts directly against material flow, inclined systems must balance complex forces acting at angles to the direction of transport. This creates a distinctive flow regime where particles experience varying degrees of suspension and sliding contact with the pipe bottom, depending on the inclination angle, air velocity and material properties. System designers must carefully consider these factors when selecting between dilute-phase operation with high air velocities or dense-phase conveying with lower velocities but higher material concentrations, as the optimal approach varies significantly with inclination angle.

Zhang et al. [6] investigated three distinct flow patterns including dispersed flow, reverse flow and half-ring flow observed under varying airflow rates in an inclined conveying pipe, as shown in Fig. 12.6, employing a combination of experimental techniques and DEM approach to analyze particle dynamics and electrostatic effects. These results reveal that at high airflow rates of 1600 L/min, particles exhibit dispersed flow, characterized by uniform distribution and forward motion. At intermediate flow rates of 1100 L/min, reverse flow emerges, with particles near the pipe center moving forward while those at the bottom slide backward, creating distinct dense and dilute regions. Further reducing the airflow rate to 1000 L/min leads to half-ring flow, where particles adhere to the pipe wall, forming an annular structure with stagnant regions at the bottom. Electrostatic forces are found to play a critical role in shaping these flow patterns. The study demonstrates that particle–wall collisions generate electrostatic charges, which influence particle adhesion and clustering, particularly in reverse and half-ring flows. The dynamic analysis of forces acting on particles, including drag, gravity, electrostatic and friction, shows that drag dominates in dispersed flow, while electrostatic forces are pivotal in reverse and half-ring flows.

Kuang et al. [7] conducted a numerical study on gas–solid flow and energy dissipation in inclined pneumatic conveyor shown in Fig. 12.7 using CFD-DEM approach. The study aims to understand the complex flow behaviors, pressure drop and energy dissipation mechanisms in inclined pipelines. Results show that the pressure drop exhibits a nonlinear trend with inclination angle, initially increasing to a peak around 50° before declining as the pipe approaches vertical. This behavior results from competing effects: at lower angles, particle–wall friction dominates, while at steeper inclinations, gravitational forces become more significant. Inclined pipes generally show higher pressure drops than horizontal or vertical configurations, with the maximum occurring at intermediate slopes. Three key forces govern pressure drop: particle–fluid interactions, gravitational effects and particle–wall friction. Below the critical angle of about 50°, wall friction is the primary resistance, while above it, gravity takes precedence. The study identifies distinct flow regimes including slug, stratified and dispersed regimes, each with characteristic pressure and energy profiles. Slug flow, occurring at low velocities, shows intense particle collisions and high energy dissipation. Stratified flow features particle settling with asymmetrical velocity distributions, while dispersed flow at high velocities exhibits uniform particle distribution and lower energy loss. Energy dissipation arises from multiple mechanisms. Particle–fluid interactions dominate in stratified and dispersed flows, while collision losses peak in slug regimes. Fluid-wall viscous dissipation becomes significant at high gas velocities. Notably, the minimum energy dissipation occurs at a lower gas velocity of about 27 m/s than the minimum pressure drop of about 34 m/s, highlighting a trade-off in system optimization. Higher solids flow rates amplify pressure drop and energy loss, particularly at intermediate inclinations.

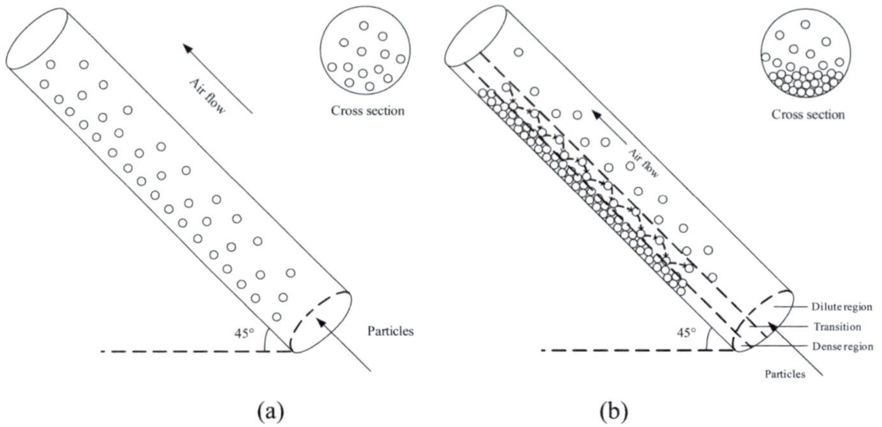

Fig. 12.6 Schematic of particle flow in three different flow patterns in pneumatic conveying: **a** dispersed flow; **b** reverse flow; and **c** half-ring flow. Panel **d** shows unsteady reverse flow with pulsating wave [6]

Fig. 12.7 Schematic illustration of forces acting on a particle assembly in inclined pneumatic conveyor [7]

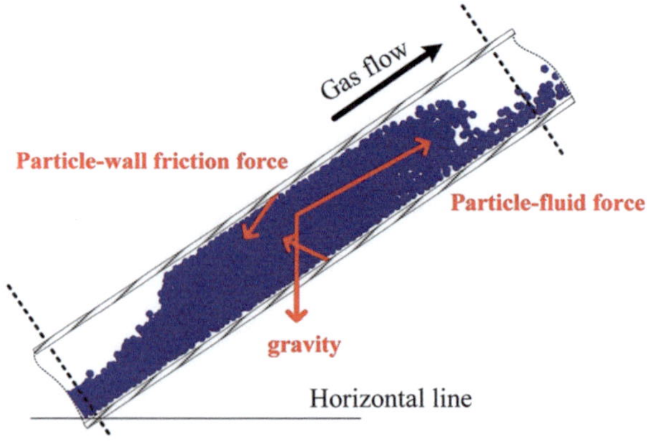

12.4 Centrifugal Aerodynamic System

Cyclone separators represent a critical class of centrifugal aerodynamic systems widely employed for gas–solid separation across industrial applications. These devices utilize inertial separation principles, where a swirling gas stream generates centrifugal forces to segregate particulate matter from the carrier gas phase. The fundamental operating mechanism involves the tangential introduction of a particle-laden gas stream into a cylindrical or conical chamber, creating a strong vortex flow. In this flow field, particles migrate radially outward due to centrifugal acceleration, subsequently descending along the cyclone walls under gravity for collection, while the cleaned gas exits through a central vortex finder. This robust separation technology offers distinct advantages, including simple construction, absence of moving parts and adaptability to high-temperature and high-pressure conditions, making it indispensable in industries ranging from power generation to chemical processing.

Existing research on cyclone separators primarily focuses on three interconnected domains: performance optimization, operational efficiency enhancement and application expansion. Many CFD-DEM numerical investigations aim to optimize geometric parameters such as inlet dimensions, vortex finder diameter and cone angle to maximize separation efficiency while minimizing pressure drop. Particular attention is also given to improving the collection efficiency of fine particulate matter such as PM2.5, which remains a persistent challenge due to the inherent limitations of centrifugal separation for submicron particles. Recent innovations in this area include the integration of electrostatic precipitation mechanisms and the development of hybrid systems combining cyclonic separation with fabric filtration.

Chu et al. [8] conducted a comprehensive study on the effects of particle size and solids loading in dense medium cyclones shown in Fig. 12.8 using CFD-DEM approach. The research focuses on understanding how particles of different sizes influence the flow patterns and performance of dense medium cyclones, which are widely used in mineral and coal processing industries for particle separation. The findings reveal that particle size significantly affects both the particle trajectories and the medium flow in the cyclone. Smaller particles, dominated by drag forces, tend to follow the fluid flow, while larger particles, influenced more by pressure gradient forces, exhibit distinct trajectories, often leading to misplacement in the overflow or underflow. Notably, particles larger than 40 mm or smaller than 0.25 mm show poor separation efficiency, with larger particles misreporting to the overflow and finer particles exhibiting reduced classification performance. The study also highlights that the spatial distribution of particle–fluid interaction forces varies with particle size, leading to different impacts on the fluid flow, such as changes in pressure drop, medium split and medium differential. Furthermore, the effect of particle top size in multisized particle systems is explored, demonstrating that the overall behavior is a collective result of individual particle size effects. The study provides insights into optimizing dense medium cyclone performance by suggesting an ideal particle size range of 0.5–40 mm for efficient separation.

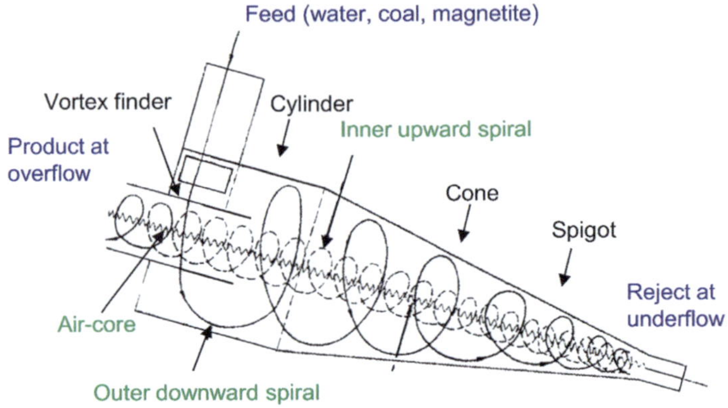

Fig. 12.8 A schematic presentation of a dense medium cyclone [8]

Wang et al. [9] investigated the effect of inlet angle on the performance of a cyclone separator shown in Fig. 12.9 using CFD-DEM approach. By varying the inlet angle ranging from −30° to 30°, the research examines the distributions of pressure drop, tangential and axial velocities and turbulent kinetic energy. The results reveal that the flow pattern consists of two distinct regions: a loss-free vortex region and a forced vortex region. Negative inlet angles are found to produce larger pressure drops compared to positive angles, while the tangential velocity remains relatively insensitive to inlet angle changes, except for larger negative angles. The axial velocity, however, increases and becomes more symmetrical with higher inlet angles. The impact of inlet angle and particle size on separation efficiency is also explored. Trajectories of particles with diameters ranging from 0.05 to 5 μm are analyzed, showing that negative inlet angles lead to short-circuiting flow, reducing separation efficiency. In contrast, an inlet angle of 15° yields the highest separation efficiency, peaking at 78%. The efficiency is further influenced by particle size, with larger particles exhibiting higher separation rates due to their greater centrifugal forces.

Fig. 12.9 Trajectories of representative particles inside cyclone separators with different inlet angles [9]

References

1. E.W.C. Lim, C.H. Wang, A.B. Yu, Discrete element simulation for pneumatic conveying of granular material. AIChE J. **52**(2), 496–509 (2006). https://doi.org/10.1002/aic.10645
2. G. Wang, S. Yang, X. Zhang, H. Wang, CFD-DEM investigation of slug dynamics in the vertical pneumatic conveying process. Chem. Eng. J. **496**, 473 (2024). https://doi.org/10.1016/j.cej.2024.153473
3. D. Yang, B. Xing, J. Li, Y. Wang, N. Hu, S. Jiang, Experiment and simulation analysis of the suspension behavior of large (5–30 mm) nonspherical particles in vertical pneumatic conveying. Powder Technol. **354**, 442–455 (2019). https://doi.org/10.1016/j.powtec.2019.06.023
4. S. Watano, S. Saito, T. Suzuki, Numerical simulation of electrostatic charge in powder pneumatic conveying process. Powder Technol. **135–136**, 112–117 (2003). https://doi.org/10.1016/S0032-5910(03)00159-1
5. O. Erken, P. Gupta, J.Y. Ooi, K.J. Hanley, Assessment of coarse grained CFD–DEM modelling strategies in horizontal plug flow pneumatic conveying. Powder Technol. **461**, 15 (2025). https://doi.org/10.1016/j.powtec.2025.121015
6. Y. Zhang, E.W.C. Lim, C.H. Wang, Pneumatic transport of granular materials in an inclined conveying pipe: comparison of computational fluid dynamics-discrete element method (CFD-DEM), electrical capacitance tomography (ECT), and particle image velocimetry (PIV) results. Ind. Eng. Chem. Res. **46**(19), 6066–6083 (2007). https://doi.org/10.1021/ie061304i
7. S.B. Kuang, R.P. Zou, R.H. Pan, A.B. Yu, Gas-solid flow and energy dissipation in inclined pneumatic conveying. Ind. Eng. Chem. Res. **51**(43), 14289–14302 (2012). https://doi.org/10.1021/ie301894d

8. K.W. Chu et al., Understand solids loading effects in a dense medium cyclone: effect of particle size by a CFD-DEM method. Powder Technol. **320**, 594–609 (2017). https://doi.org/10.1016/j.powtec.2017.07.032
9. S. Wang, H. Li, R. Wang, X. Wang, R. Tian, Q. Sun, Effect of the inlet angle on the performance of a cyclone separator using CFD-DEM. Adv. Powder Technol. **30**(2), 227–239 (2019). https://doi.org/10.1016/j.apt.2018.10.027

Chapter 13
Applications to Powder Transport Systems

13.1 Screw Conveyor

Screw conveyors, also known as auger conveyors, are mechanical systems widely used to transport bulk materials such as powders, granules and pellets. They consist of a helical screw blade (flight) rotating in a tube or trough, which moves materials along the axis of the screw. Due to their simplicity, versatility and ability to handle a wide range of materials, from fine powders to coarse aggregates, screw conveyors are essential in industries such as agriculture, food processing, mining, chemical manufacturing and recycling. Their enclosed design minimizes dust emissions and contamination, making them suitable for handling sensitive or hazardous materials. Additionally, screw conveyors can be configured horizontally, vertically or at an incline, allowing flexible integration into various processing lines. The efficiency of a screw conveyor depends on factors such as screw geometry, rotational speed, material properties and operational conditions. Poor design or improper operation can lead to issues like particle degradation, uneven flow, excessive wear or energy inefficiency. Therefore, optimizing screw conveyors is crucial for maintaining productivity, reducing maintenance costs and ensuring consistent material handling.

13.1.1 Research Focus

The DEM has become a powerful tool for analyzing the complex dynamics of particles in screw conveyors. Unlike traditional analytical or experimental methods, DEM provides detailed insights into particle-scale interactions, enabling engineers to visualize and quantify phenomena such as particle flow patterns, force distributions and wear mechanisms. Figure 13.1 displays the DEM research framework for screw conveyors. Below are the key aspects of DEM-based screw conveyor research:

R. K. Agarwal and Y. Shao, *Theory and Application of Discrete Element Method to Simulation of Multiphase Flows*, https://doi.org/10.1007/978-3-032-11157-9_13

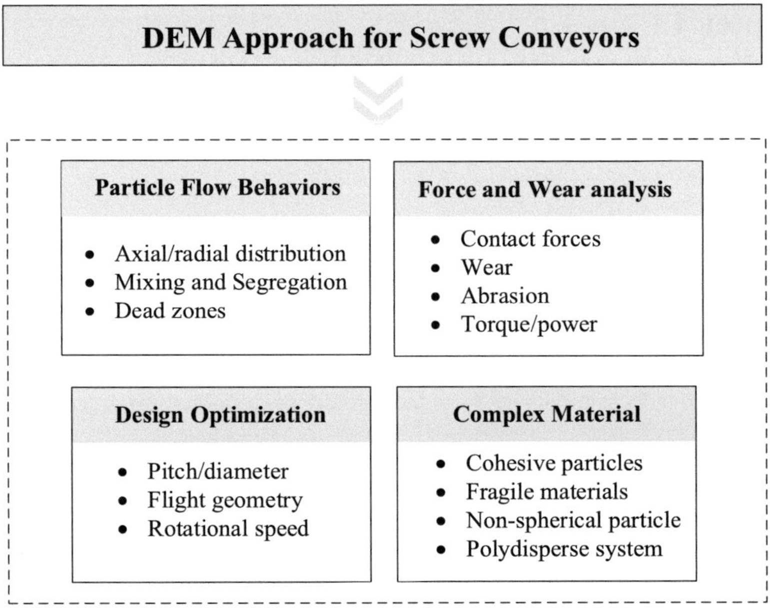

Fig. 13.1 DEM research framework for screw conveyors

(1) Particle Flow Behaviors

DEM simulations allow to study how particles move along the screw conveyor, including their axial and radial distribution. By tracking individual particles, DEM reveals segregation tendencies, especially in mixtures of different sizes or densities. For instance, smaller or denser particles may migrate toward the bottom of the trough, while lighter particles are pushed upward by the screw's motion. Additionally, DEM helps identify stagnant zones where material accumulates, leading to blockages or inefficient transport. Understanding these flow patterns is critical for optimizing screw design to ensure uniform material discharge and prevent operational disruptions.

(2) Forces and Wear Analysis

Mechanical wear is also one of the primary concerns in screw conveyor operation, particularly on the screw flight and trough surfaces. DEM simulations quantify contact forces between particles and the conveyor structure, helping predict high-stress regions prone to abrasion. This information is valuable for selecting wear-resistant materials or modifying screw geometry to distribute forces more evenly. Furthermore, DEM can estimate the torque required to drive the screw, which directly impacts power consumption. By analyzing how particle properties such as hardness and shape influence these forces, engineers can design more energy-efficient and durable systems.

(3) Influence of Screw Design Parameters

DEM enables parametric studies to evaluate how different screw configurations affect performance. Key design variables include screw pitch and diameter, flight geometry, rotational speed and trough design. Larger pitches may increase throughput but reduce control over fine materials, while smaller pitches enhance mixing but require higher torque. Variations such as ribbon flights, cut flights or tapered screws can be tested to optimize flow for specific materials. DEM helps determine the optimal speed to balance throughput and particle damage, especially for fragile materials like grains or pharmaceuticals. U-trough, tubular or enclosed designs influence particle confinement and flow behavior. By simulating these parameters, DEM reduces the need for costly trial-and-error prototyping and accelerates the development of customized screw conveyors.

(4) Interaction with Complex Materials

Many industrial materials exhibit non-ideal behaviors, such as cohesion, abrasiveness or irregular shapes. DEM can incorporate advanced particle models to account for these properties. For example, DEM simulations using superquadric or polyhedral shapes better represent real-world materials like biomass; adhesion models simulate sticky materials like moist clay or food pastes that tend to clump or adhere to the screw surface; DEM can predict fragmentation of brittle materials, helping minimize degradation during transport.

13.1.2 Example Applications

Owen and Cleary [1] employed the DEM approach to analyze the performance of screw conveyors shown in Fig. 13.2 under varying operational conditions, focusing on rotational speed, inclination angle and fill level. The DEM model simulates a single-pitch screw conveyor with periodic boundary conditions, using spherical particles to represent granular material. Simulations explore a range of rotational speeds (600, 1000 and 1400 rpm), inclination angles (0°–90° in 10° increments) and fill levels (30, 50 and 70% by volume), as shown in Fig. 13.3. The performances examined include particle velocities, mass flow rates, power consumption and energy dissipation mechanisms.

The research identifies two distinct flow patterns based on screw inclination. At low angles, particles form a recirculating heap, with avalanching flow down the free surface and vertical transport along the screw face. This regime is characterized by size segregation and high axial transport efficiency. In contrast, vertical conveyors exhibit a uniform particle bed on the screw blade, where centrifugal forces dominate, and axial transport diminishes. Intermediate angles show a superposition of these flow patterns, with the transition from heap-dominated to bed-dominated behavior occurring around 30°–60° inclination. The study quantifies particle dynamics, revealing that axial velocity decreases linearly with inclination, while swirl velocity increases,

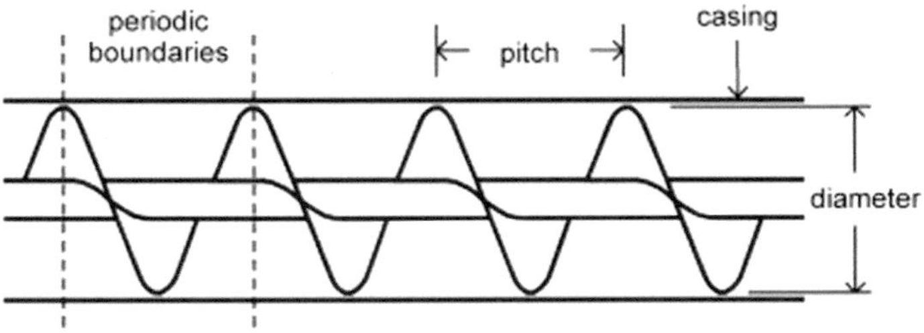

Fig. 13.2 Single flight screw conveyor (pitch equal to diameter) [1]

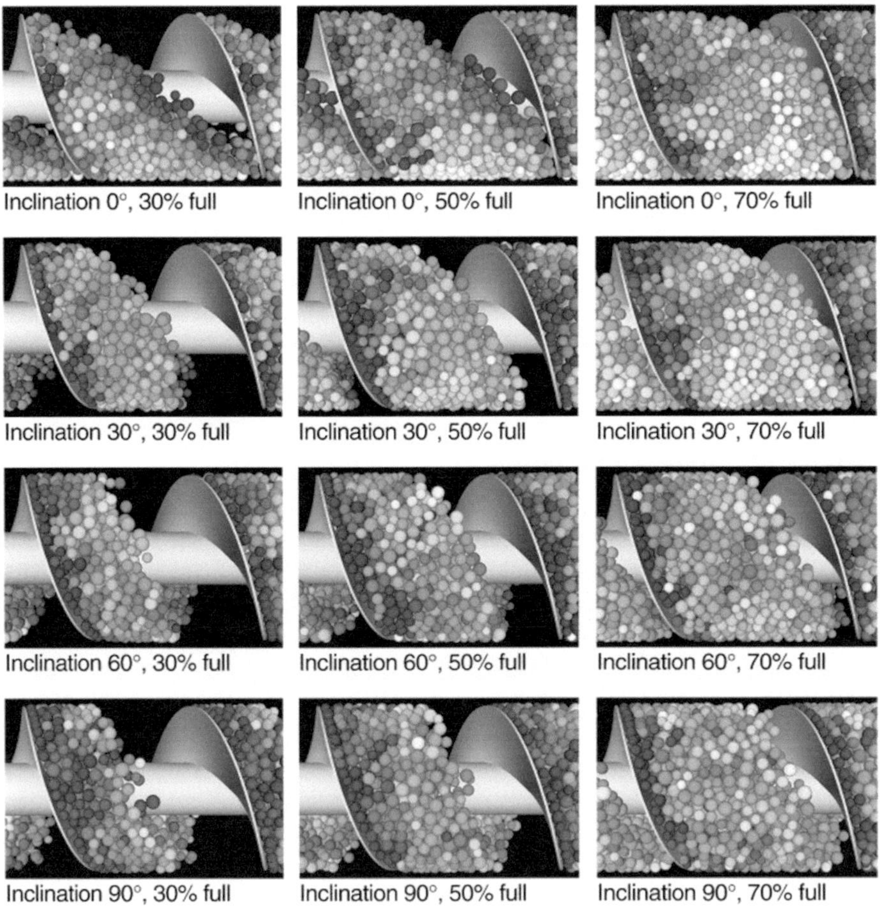

Fig. 13.3 Particle distributions within the screw conveyor inclined at various angles and operating with different fill levels. Particles are colored by their speed: from light gray to dark gray for 0.4–0.9 m/s [1]

peaking at high angles due to centrifugal effects. Energy and power analysis highlights the trade-offs between transport efficiency and operational costs. Power draw rises with inclination and fill level, driven by the energy required to sustain particle swirl motion. Energy dissipation occurs primarily through shear in high-inclination flows and impacts in low-inclination or high-fill conditions. Collision energy spectra reveal that modal collision energy increases with inclination and screw speed, with shear dissipation dominating in bed-flow regimes.

Sun et al. [2] investigated the transport behavior of non-spherical particles in horizontal screw conveyors shown in Fig. 13.4 using the DEM. A superellipsoid model to represent particles of varying shapes is employed, from near-spherical to cubic, and incorporates a shear impact energy model to predict conveyor wear. The research aims to provide a comprehensive understanding of how particle shape influences flow dynamics, energy dissipation and equipment wear. The results reveal that spherical particles exhibit higher axial and angular velocities compared to non-spherical particles, which experience hindered motion due to interlocking and block effects. The flow patterns show distinct recirculation behavior, with particles moving upward along the screw blade and avalanching downward, though this pattern weakens as particle shape becomes more cubic. Contact force analysis indicates that non-spherical particles generate larger forces, especially at higher feeding rates, leading to greater energy dissipation and wear. The mass flow rate analysis demonstrates that while spherical particles achieve higher initial throughput, the difference diminishes at steady state, with non-spherical particles showing reduced flow rates at high feeding densities due to accumulation effects. Additionally, non-spherical particles exhibit higher energy dissipation in particle–wall collisions, increasing the risk of particle breakage and equipment wear. The SIEM predictions show maximum wear occurring at the screw blade edges, where impact forces are concentrated. Wear rates increase with feeding rate and particle shape index but decrease with higher rotational speeds, suggesting operational trade-offs between throughput and equipment longevity.

Ge et al. [3] conducted DEM simulations on the granular flow and heat transfer behaviors in an industrial-scale screw conveyor, as shown in Fig. 13.5. The screw conveyor consists of upper and lower spirals, where particles are heated in the upper spiral and preserved in the lower spiral. The results reveal that the average particle velocities in the upper and lower spirals are 0.3908 and 0.2620 m/s, respectively. Particle segregation due to size differences is noted, with smaller particles concentrating near the walls and some adhering to the spiral blades. Stagnation zones at the conveyor ends are identified, which may lead to localized slagging. As for the heat transfer characteristics, the upper spiral primarily has heated particles, raising their average temperature to 624.37 K at the exit. The lower spiral acts as a heat preservation zone, maintaining particles at ~673.15 K.

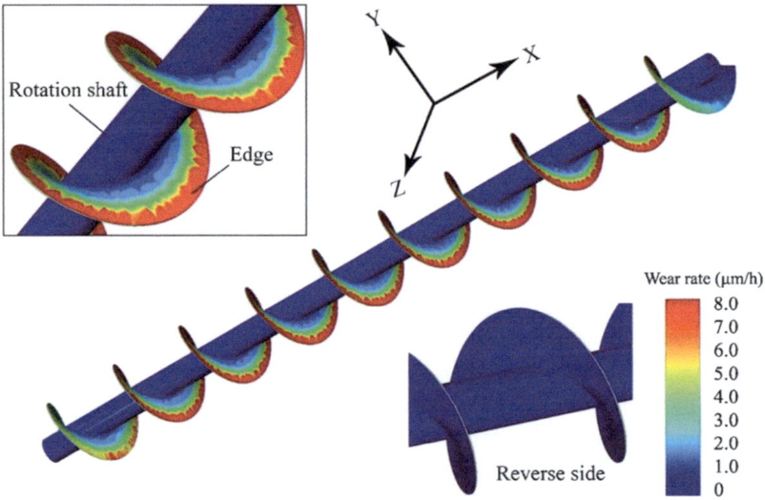

Fig. 13.4 Distribution of the wear rate on the screw blade at $q = 6000$ particles/s and $\Omega = 60$ rpm [2]. Note that the shaft is colored by the degree of erosion

13.2 Screw Feeder

A screw feeder is a specialized device designed to meter and control the discharge of bulk materials from a hopper or storage bin into a downstream process, such as a reactor, mixer or pneumatic conveying system, as shown in Fig. 13.6. Unlike a screw conveyor, which primarily transports materials over a distance, a screw feeder operates under flood-loaded conditions to ensure a consistent and controlled discharge rate. Key design differences include variable pitch screws, tapered flights and slower rotational speeds, which help prevent material flushing and ensure uniform withdrawal from the hopper. This distinction makes screw feeders critical in applications requiring precise feed control, such as in chemical processing, food production and pharmaceutical manufacturing.

13.2.1 Research Focus

When applying the DEM to investigate screw feeders, researchers concentrate on several critical aspects to optimize performance and reliability. One primary focus is flow dynamics and discharge control, where DEM simulations help identify issues like ratholing, bridging and erratic flow caused by material properties or screw design. By analyzing particle trajectories and forces, DEM can reveal how variable pitch screws or tapered flights influence flow stability, particularly for cohesive or size-varying materials. Another key area of study is particle–wall and particle–particle

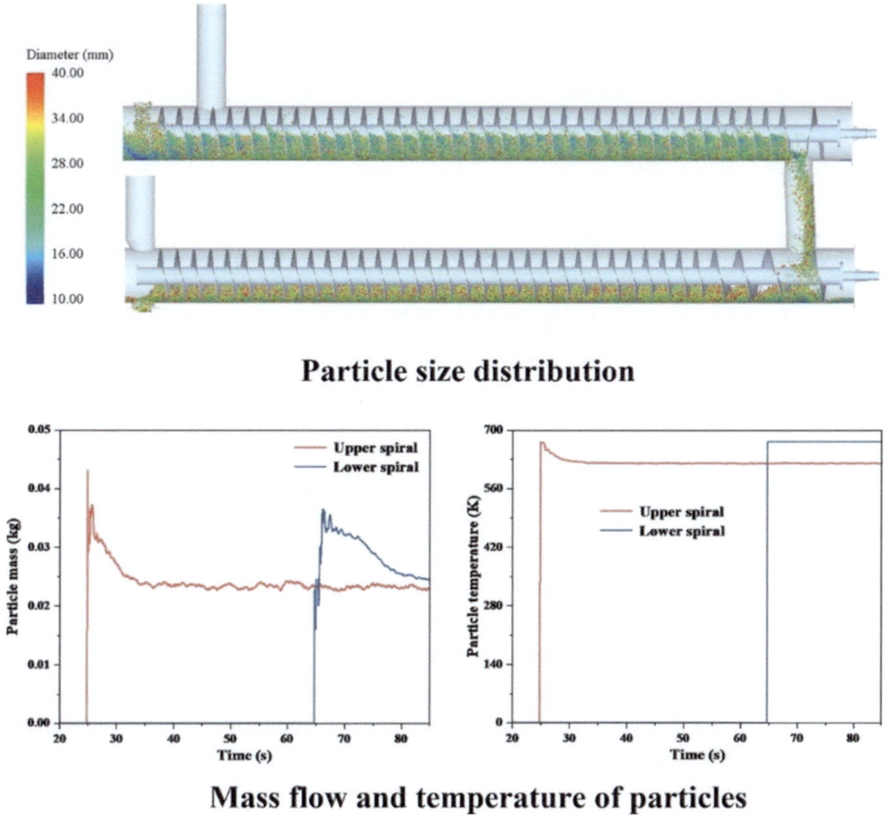

Fig. 13.5 Granular flow and heat transfer in an industrial-scale screw conveyor [3]

interactions, which significantly affect feeder performance. DEM allows to examine friction, cohesion and adhesion forces that may disrupt material flow. Additionally, DEM simulations can predict wear patterns on screw flights and troughs, especially when handling abrasive materials. Feed rate optimization is also a major research focus, with DEM used to correlate screw speed, fill level and discharge rate for different bulk solids. This helps in calibrating feed rate models and ensuring consistent performance across varying operating conditions. Furthermore, DEM aids in assessing power consumption and torque requirements, which are crucial for motor sizing and energy efficiency. Start-up torque analysis is particularly important for cohesive materials, where sudden load changes can strain the system. Finally, DEM is helpful in evaluating specialized feeder designs, such as twin-screw feeders for handling difficult materials or flexible screw feeders for fragile particles. By simulating these configurations, researchers can optimize screw geometry, hopper design and integration with other processes to enhance overall system performance.

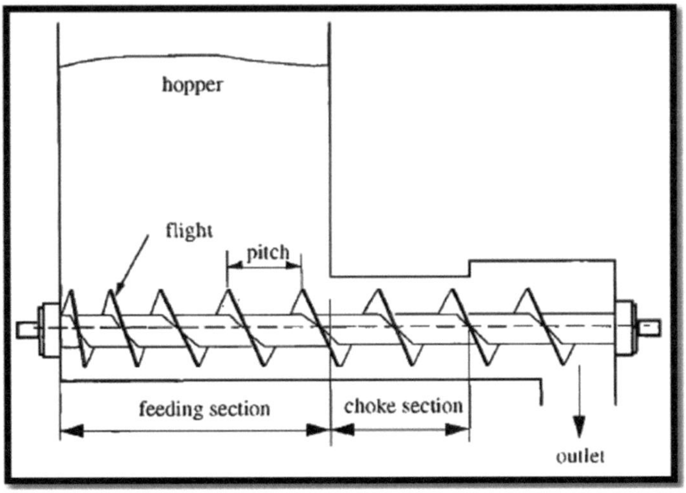

Fig. 13.6 Screw feeder along with hopper [4]

13.2.2 Example Applications

Fernandez et al. [5] investigated the effect of screw design on hopper drawdown in horizontal screw feeders using DEM. The research focuses on six screw designs including variations in pitch, flight diameter and core taper shown in Fig. 13.7, to evaluate their impact on particle flow uniformity, power consumption and abrasive wear. The study reveals that screw design significantly influences the uniformity of material drawdown from the hopper. Screws with variable pitch and tapered cores (Screw F) achieve the most uniform drawdown, while standard designs (Screw A) exhibit preferential drawdown near the hopper rear, leading to stagnant zones. The DEM simulations identify two primary particle pathways: a dominant surface-to-rear flow and a slower bed deformation flow. A mechanism of particle compaction in the screw is observed, where shear-induced densification progressively increases packing density along the screw length. This compaction affects the available transport volume and drawdown uniformity. Power draw varies significantly among screw designs, with Screw B (tapered flight) exhibiting the lowest consumption and Screw A (standard design) the highest. Abrasive wear is highest at the screw exit and on outer blade edges due to increased particle forces and slip velocities. Designs with blade heights less than one particle diameter showed reduced wear but lower transport efficiency. Reducing screw surface friction increases mass flow rates but slightly compromises drawdown uniformity, while higher friction reduces flow rates and increases power consumption.

Hou et al. [6] employed the DEM to investigate the flow of cohesive particles in screw feeders shown in Fig. 13.8, focusing on the effects of interparticle cohesion, screw speed, screw length and container design. The simulations identify three

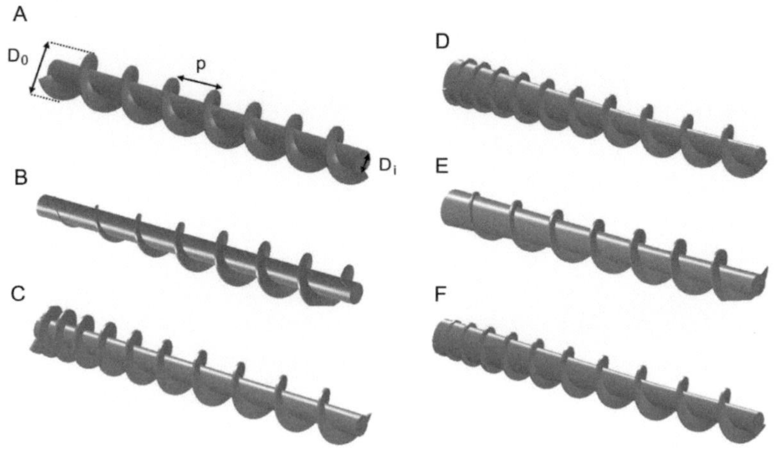

Fig. 13.7 Screw design variations. Screw **a**: standard; screw **b**: taper flight; screw **c**: variable pitch; screw **d**: variable pitch and taper; screw **e**: tapered flight and core with constant pitch; screw **f**: optimal flight, tapered core and variable pitch [5]

distinct flow regimes based on cohesive force magnitude: continuous flow, intermittent flow and stable arch flow, as shown in Fig. 13.9. Continuous flow occurs at low cohesion, where particles move smoothly through the feeder. Intermittent flow arises at moderate cohesion, characterized by unstable particle movement due to partial force network formation. At high cohesion, particles form stable arches, completely blocking flow. A correlation is derived to predict flow rates as a function of cohesive force and screw speed, offering a practical tool for performance estimation in industrial settings. The study also examines the impact of screw length in the charging container. Longer screws alter drawdown patterns and enhance perturbations, which disrupt force networks and delay the onset of stable arch flow. An optimal screw length of three pitches is identified, balancing flow rate and energy consumption. Additionally, the design of the container is found to significantly influence flow dynamics. While both bin and hopper designs exhibit similar flow regimes, the hopper's converging walls exacerbate cohesion effects by squeezing particles near the outlet, leading to stronger agglomerates and reduced flow rates compared to the bin.

Lian et al. [7] investigated the mixed feeding process of coal and cylindrical biomass particles in a screw feeder using the DEM, as shown in Fig. 13.10. The research focuses on understanding how key operational parameters including biomass feeding ratio, feeding rate and screw rotational speed affect the feeding performance, particularly in terms of continuity, uniformity and stability. The coal particles are modeled as spherical, while the cylindrical biomass particles (pine wood) are constructed using the multisphere model with seven sphere elements. The simulations are conducted on a shaftless horizontal screw feeder with a pitch of 23.5 mm and a diameter of 35 mm, varying parameters such as biomass feeding ratio (0–100%), feeding rate (63–183 particles/s) and screw rotational speed (15–35 rpm).

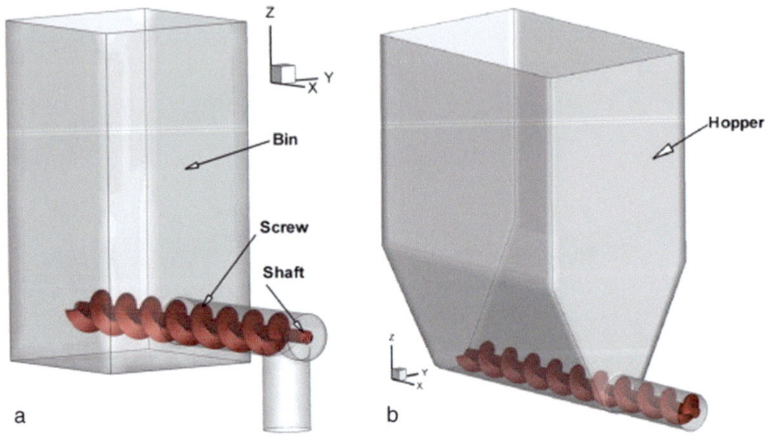

Fig. 13.8 Schematic of screw feeders with different container designs: **a** bin and **b** hopper [6]

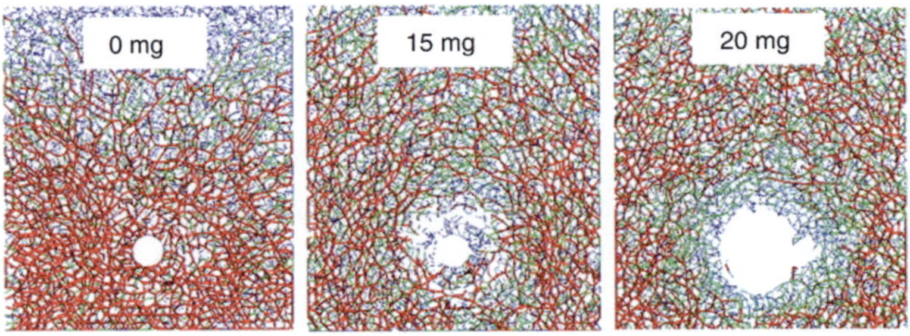

Fig. 13.9 Spatial distribution of normal contact forces at different cohesive forces (0, 15 and 20 mg) where the rotational speed is 600 rpm [6]

The findings reveal that the biomass feeding ratio significantly influences the stability of the feeding process. Higher biomass feeding ratios ($\geq$70%) result in more stable real-time mass flow rates and biomass blending ratios, reducing fluctuations. In contrast, lower ratios ($\leq$30%) lead to dispersed biomass distribution and increased variability in blending. Despite the differences in particle density and shape, the mixture remains uniform along the screw casing, with no significant segregation observed. This uniformity ensures consistent feeding, which is crucial for applications like co-firing in reactors. The study also highlights the impact of feeding rate, where low rates cause irregular blending due to sparse particle distribution, while high rates increase mass flow fluctuations and risk biomass particle aggregation near the hopper wall, potentially hindering flow. An optimal feeding rate, such as B103C103 (103 particles/s for both coal and biomass), balances stability and throughput. Screw rotational speed plays a critical role in determining feeding stability. Higher rotational speeds ($\geq$25 rpm) reduce variability in both mass flow rate and blending ratio,

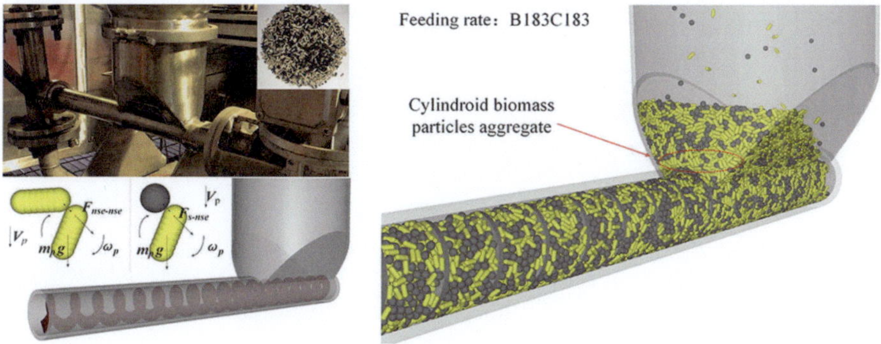

Fig. 13.10 Mixed feeding process of coal and cylindroid biomass particles in a screw feeder [7]

promoting smoother operation. However, lower speeds of 15 rpm can lead to particle stacking and deviations in the blending ratio, compromising process efficiency.

References

1. P.J. Owen, P.W. Cleary, Prediction of screw conveyor performance using the discrete element method (DEM). Powder Technol. **193**(3), 274–288 (2009). https://doi.org/10.1016/j.powtec.2009.03.012
2. H. Sun, H. Ma, Y. Zhao, DEM investigation on conveying of non-spherical particles in a screw conveyor. Particuology **65**, 17–31 (2022). https://doi.org/10.1016/j.partic.2021.06.009
3. Y. Ge, J. Yu, J. Lin, S. Wang, K. Luo, J. Fan, DEM simulation of granular flow and heat transfer in an industrial-scale screw conveyor. Powder Technol. **452**, 537 (2025). https://doi.org/10.1016/j.powtec.2024.120537
4. D. Minglani, A. Sharma, H. Pandey, R. Dayal, J.B. Joshi, S. Subramaniam, *A Review of Granular Flow in Screw Feeders and Conveyors* (Elsevier, 2020). https://doi.org/10.1016/j.powtec.2020.02.066
5. J.W. Fernandez, P.W. Cleary, W. McBride, Effect of screw design on hopper drawdown of spherical particles in a horizontal screw feeder. Chem. Eng. Sci. **66**(22), 5585–5601 (2011). https://doi.org/10.1016/j.ces.2011.07.043
6. Q.F. Hou, K.J. Dong, A.B. Yu, DEM study of the flow of cohesive particles in a screw feeder. Powder Technol. **256**, 529–539 (2014). https://doi.org/10.1016/j.powtec.2014.01.062
7. G. Lian, W. Zhong, X. Liu, DEM study on the mixed feeding process of coal and cylindroid biomass particles in a screw feeder. Adv. Powder Technol. **32**(7), 2543–2554 (2021). https://doi.org/10.1016/j.apt.2021.05.031

Chapter 14
Applications to Pumps

A pump is a mechanical device designed to move fluids from one place to another by converting mechanical energy into hydraulic energy. Pumps are essential in various industries, including water supply, oil and gas, chemical processing, agriculture and wastewater management. They work by creating a pressure difference, forcing the fluid to flow from a region of low pressure to high pressure. Depending on the application, pumps may handle clean water, corrosive chemicals, viscous oils or slurries containing solid particles. The efficiency, durability and performance of a pump depend on its design, working principle, and the nature of the fluid being transported.

Pumps can be broadly classified into two main categories based on their working principle: dynamic pumps and positive displacement pumps. Dynamic pumps, also known as kinetic pumps, operate by imparting velocity to the fluid, which is then converted into pressure. The most common type is the centrifugal pump, which uses a rotating impeller to accelerate the fluid outward, creating a centrifugal force that moves it through the system. Other variations include axial flow pumps, where fluid moves parallel to the pump shaft, and mixed flow pumps, which combine radial and axial flow characteristics. These pumps are widely used in water distribution, irrigation and industrial processes due to their ability to handle large flow rates efficiently. Positive displacement pumps, on the other hand, move fluid by trapping a fixed volume and forcing it into the discharge pipe. These pumps are further divided into reciprocating pumps such as piston and diaphragm pumps and rotary pumps like gear, screw and vane pumps. Reciprocating pumps use a back-and-forth motion to displace fluid, making them suitable for high-pressure applications like hydraulic systems and oil drilling. Rotary pumps, meanwhile, rely on rotating mechanisms to transfer fluid, making them ideal for viscous liquids or precise metering in chemical dosing. Beyond these primary classifications, pumps can also be categorized based on specialized working principles, such as the hydraulic jet pump, which operates on fluid momentum transfer rather than mechanical displacement or rotational dynamics. Unlike dynamic or positive displacement pumps, hydraulic

R. K. Agarwal and Y. Shao, *Theory and Application of Discrete Element Method to Simulation of Multiphase Flows*, https://doi.org/10.1007/978-3-032-11157-9_14

jet pumps utilize a high-velocity motive fluid to entrain and pressurize a secondary fluid through the Venturi effect, making them ideal for applications requiring robust handling of abrasive, corrosive or gas-entrained fluids. Additionally, pumps can also be categorized by their application, installation type or drive mechanism. Each type has unique advantages tailored to specific operational needs, ensuring optimal performance across different engineering and industrial scenarios.

14.1 Research Focus

CFD-DEM approach has emerged as a powerful numerical approach for studying complex fluid-particle interactions in pumps. This method combines CFD for fluid flow simulation with DEM for tracking individual particle dynamics, making it particularly useful for analyzing solid–liquid two-phase flows, abrasive wear and particle transport efficiency in various types of pumps. The application of CFD-DEM in pumps is especially relevant in industries dealing with slurry transport. For example, in centrifugal slurry pumps, the method helps analyze how solid particles interact with the impeller and volute casing, leading to localized wear and efficiency losses. One of the primary research focuses in CFD-DEM pump studies is wear prediction and its mitigation. Since abrasive wear is a major cause of pump failure in slurry applications, DEM is used to track particle–wall collisions and apply wear models, such as the Finnie model [1], Oka model [2], E/CRE Erosion model [3], Archard model [4] and Mclaury model [5], to predict erosion patterns on impeller blades, casings and other critical components. Detailed description of the wear models can be seen in the review article [6]. By optimizing geometric parameters such as blade angle, number of vanes or surface coatings, engineers can design more durable pumps. Another key area of investigation is solid–liquid flow dynamics, where researchers examine how particle size, shape and concentration affect pump performance. Large or dense particles may settle in low-velocity regions, leading to blockages, while fine particles may follow fluid streamlines more closely but contribute to increased hydraulic losses. CFD-DEM helps quantify these effects, enabling better pump selection and operational guidelines for different slurry conditions. Additionally, particle-induced turbulence and energy losses are critical research topics. Solid particles can disrupt fluid flow, increasing turbulence and reducing pump efficiency. By analyzing these interactions, researchers can develop improved impeller designs or operational strategies such as adjusting rotational speed or inlet conditions to minimize energy consumption. Finally, validation and model refinement remain ongoing challenges. Advances in parallel computing and coarse-graining methods are also being explored to make simulations more feasible for industrial-scale applications.

14.2 Abrasive Wear

Kang et al. [7] investigated the wear characteristics of a centrifugal pump transporting solid–liquid mixtures through a combination of experimental and numerical methods. The experiment involves a 30-h test where the pump transports a mixture of water and brown fused alumina particles with an average diameter of 2.0 mm and a solid mass fraction of 10%. The results reveal severe wear on the inlet edges of the long blades, which develops wavy profiles on the outlet parts of the blades, which are worn through. The volute tongue also exhibits significant wear, while the rest of the volute remains relatively intact. The numerical simulation employed a CFD-DEM approach to analyze the two-phase flow and wear patterns. The study explores the effects of particle diameter (1.0–3.0 mm) and solid mass fraction (5–15%) on wear. Simulation results show that wear on the long blades is significantly more severe than on the volute, with the ratio of average wear depths reaching up to 64.5 at a solid mass fraction of 10%. The blade inlet edge wear is primarily caused by normal cumulative contact energy, while the outlet part is influenced by tangential cumulative contact energy. Increasing the particle diameter leads to a 45% reduction in average wear depth but a 34% increase in maximum wear depth due to the higher kinetic energy of larger particles. The solid mass fraction is identified as a critical factor, with higher concentrations exacerbating wear, particularly for larger particles, as shown in Figs. 14.1 and 14.2. Additionally, the results also highlight the non-uniform distribution of wear on the volute, with severe erosion concentrated near the volute tongue and downstream sections.

Wang et al. [8] investigated the effects of operating conditions on the erosion characteristics of a centrifugal slurry pump shown in Fig. 14.3 using the DDPM approach. The research systematically examines the impact of key variables including slurry flow rate (Q), impeller speed (N), particle size (d_s) and solid concentration (C_s) on pump behavior, with a focus on hydraulic performance and wall erosion.

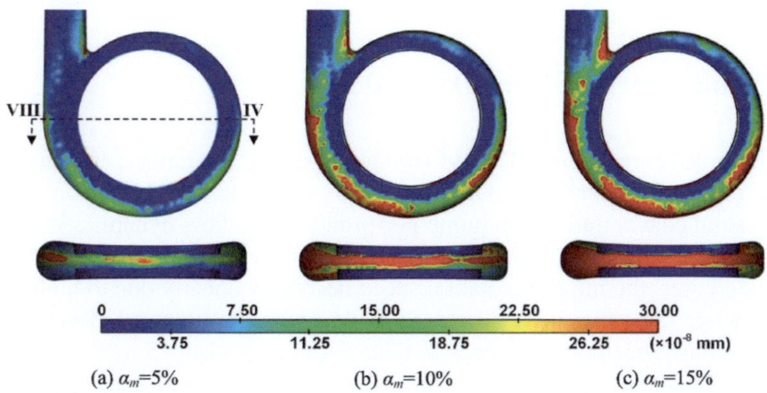

Fig. 14.1 Distribution of average wear depth over inner wall of the volute at solid mass fractions of 5, 10 and 15% [7]

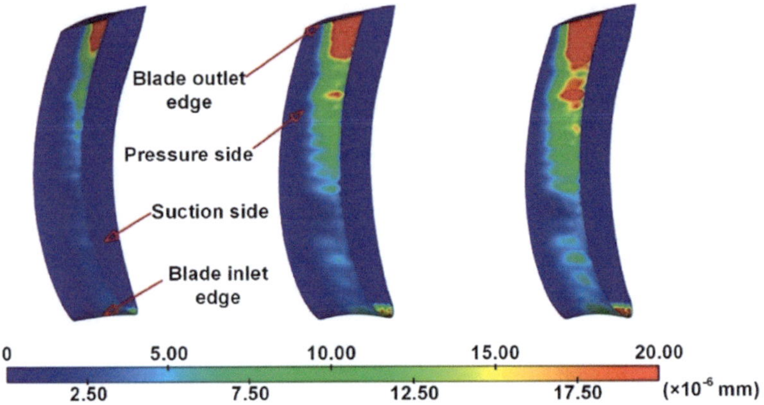

Fig. 14.2 Distribution of depth of wear over the pressure and suction sides of a long blade at solid mass fractions of 5, 10 and 15% [7]

Key findings reveal that slurry flow rate significantly influences the distribution and severity of erosion in the volute casing. Under-BEP (Best Efficiency Point) conditions, erosion is more uniformly distributed, while over-BEP conditions intensify wear in the volute belly and side liners. The total erosion rate increases with higher flow rates, particularly at elevated solid concentrations. Impeller speed enhances pump head and efficiency but exacerbates erosion, especially in the volute casing, due to increased particle kinetic energy. Particle size exhibits a U-shaped relationship with erosion, with fine particles (50–100 μm) causing severe wear in side-liner clearances and larger particles (300–700 μm) dominating volute erosion. Solid concentration amplifies these effects, with higher concentrations leading to greater erosion rates and altered efficiency trends.

14.3 Fluid-Particle Flow Dynamics

Wang et al. [9] carried out a DDPM simulation on the hydraulic performance and particle motion in centrifugal slurry pumps, as shown in Fig. 14.4. The effects of varying solid concentrations on pump performance and flow dynamics under fixed motor power conditions are investigated. The findings reveal that as solid concentration increases, the impeller rotation speed decreases linearly due to higher slurry density, while the pump head decreases, with a more pronounced effect at higher concentrations. The pump efficiency initially rises to a peak at 15 wt% solid concentration before declining, attributed to the opposing trends of flow rate and pump head contributions. The analysis of particle motion shows that particles accumulate in the volute casing, particularly in the belly region, due to centrifugal forces. Near-wall

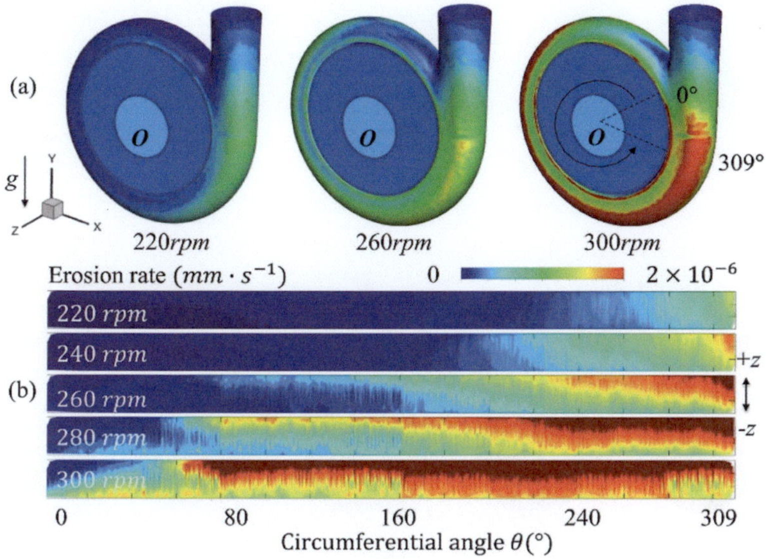

Fig. 14.3 Snapshot shows spatial distribution of erosion rate at different impeller speed (N = 220–300 rpm, $C_s = 10$ wt%) on **a** Pump **b** the Volute [8]

particle volume fractions exceed feed concentrations, indicating significant particle accretion. The highest particle velocities occur at the impeller's exterior edge, where particle volume fractions are lower, and energy transfer from the impeller accelerates downstream particles. The pre-swirl vane, designed to guide particle–fluid flow, further enhances particle acceleration toward the impeller passages. Temporal evolution of particle discharge rates indicates that higher solid concentrations lead to faster flow establishment, with fluctuations diminishing over time.

Wang et al. [10] investigated the effect of particle size on the vortex structure and flow characteristics in a semi-open centrifugal pump shown in Fig. 14.5. The focus is on understanding how particle size influences the tip leakage vortex (TLV) and other vortex structures, as well as the overall hydraulic performance of the pump. Key findings reveal that particle size significantly affects the vortex dynamics and flow stability. Smaller particles of 0.1 and 0.3 mm exhibit good following behavior with the fluid, leading to increased particle migration into the tip clearance. This results in frequent particle impacts on the TLV, causing its breakup, separation and fusion, which disrupts the vortex structure and aggravates flow instability. In contrast, larger particles of 0.5 and 0.7 mm tend to move toward the blade pressure surface and hub due to their greater inertia, reducing their interaction with the TLV and allowing the vortex structure to remain more stable. The study also identifies that the relative vorticity stretching term is a primary driver of TLV formation and development, while Coriolis force and viscous diffusion terms play significant roles in vorticity generation and dissipation.

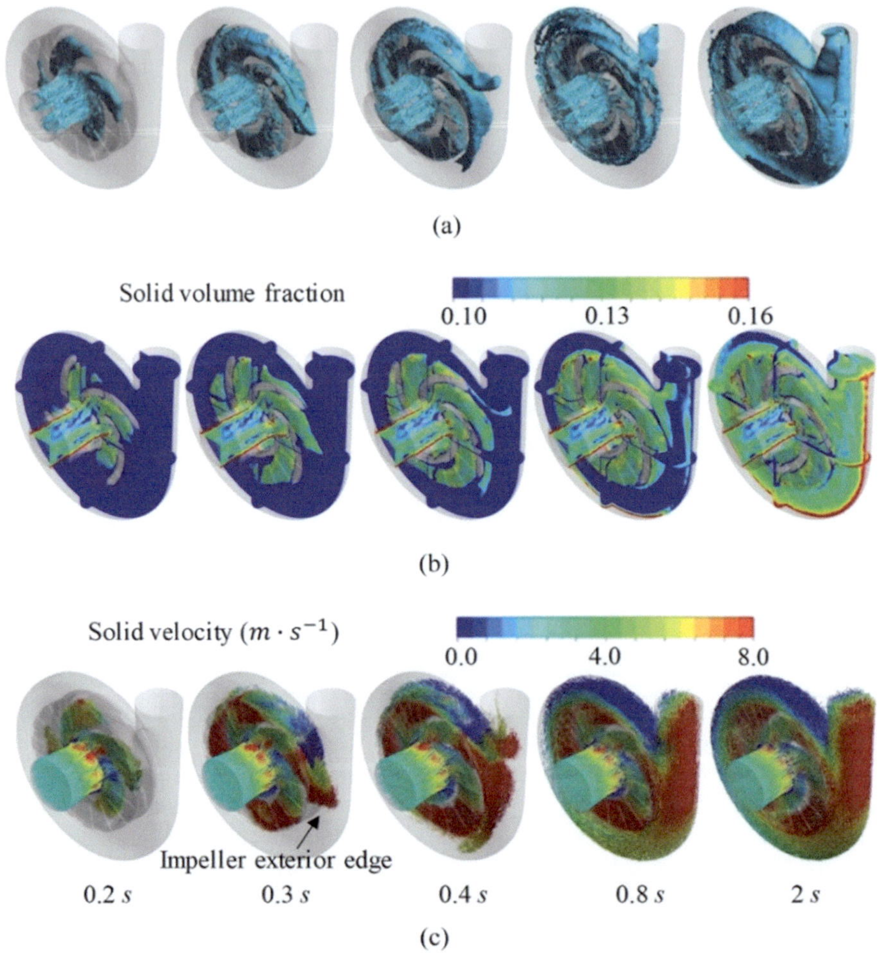

Fig. 14.4 Time-evolution of **a** isosurface (particle volume fraction $\alpha_s = 0.1$) and **b** spatial distribution of particle volume fraction, as well as **c** spatial distribution of solid velocity [9]

Wen et al. [11] presented a comprehensive study on the vortex characteristics in hydraulic jet pumps during the vertical hydraulic conveying of coarse particles, as shown in Fig. 14.6. Four distinct vortex types including fluid entraining vortex (FEV), fluid rotating vortex (FRV), fluid attacking vortex (FAV) and fluid mixing vortex (FMV) are identified, and their roles in energy dissipation and hydraulic performance are investigated. The FEV and FRV are primarily influenced by structural components like the suction port and chamber, while the FMV and FAV are strongly affected by the mixer-nozzle width, a key geometric parameter. The study reveals that the mixer-nozzle width significantly impacts vortex dynamics: increasing the width enhances the FAV area initially but reduces it at higher widths due to decreased secondary flux,

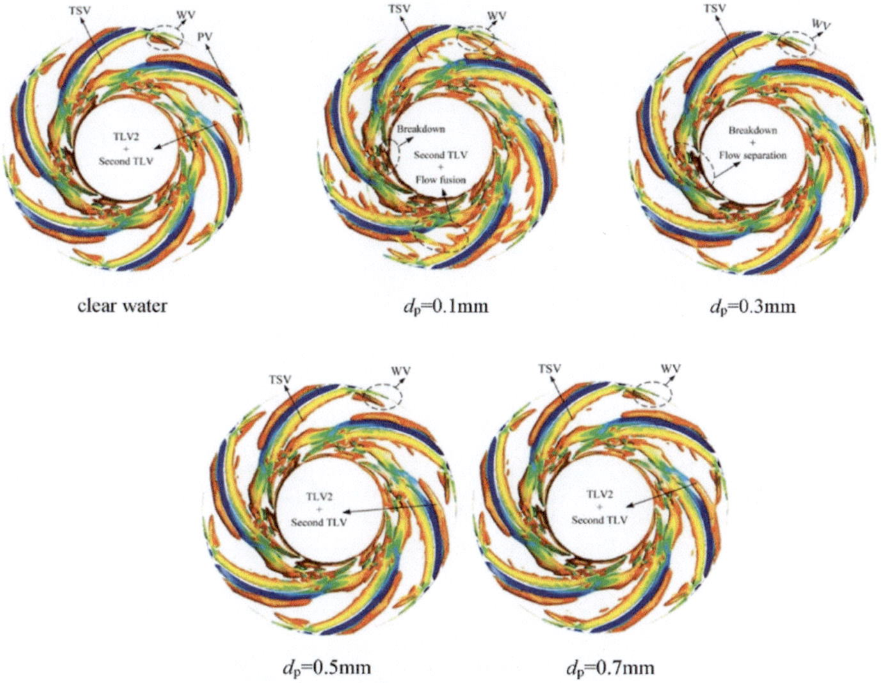

Fig. 14.5 Leakage vortex structure under different particle sizes [10]

while the FMV area diminishes sharply beyond a critical width. These vortex behaviors directly influence the hydraulic jet pump's operational regimes, with an optimal width coefficient of 2.0 identified to maximize particle conveying efficiency and minimize energy losses. Furthermore, the research highlights the transition between two jet regimes, that is Regime I (stable performance) and Regime II (performance degradation), driven by changes in FAV dominance.

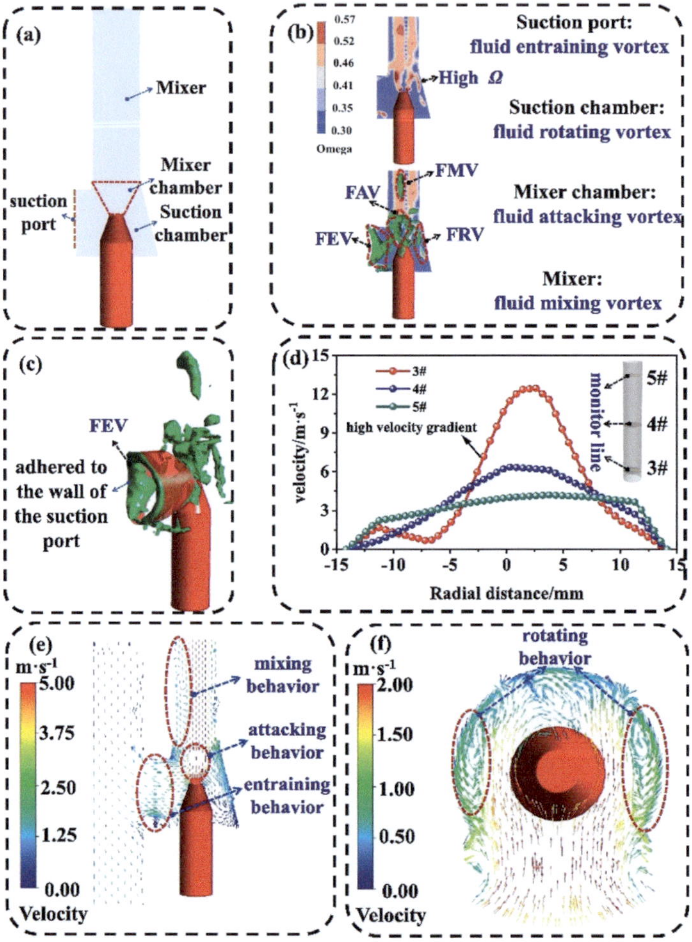

Fig. 14.6 Vortex core and fluid vector distribution in the hydraulic jet pump [11]

References

1. I. Finnie, Erosion of surfaces by solid particles. Wear **3**(2), 87–103 (1960). https://doi.org/10.1016/0043-1648(60)90055-7
2. Y.I. Oka, H. Ohnogi, T. Hosokawa, M. Matsumura, The impact angle dependence of erosion damage caused by solid particle impact. Wear **203–204**, 573–579 (1997). https://doi.org/10.1016/S0043-1648(96)07430-3
3. Y. Zhang, E.P. Reuterfors, B.S. McLaury, S.A. Shirazi, E.F. Rybicki, Comparison of computed and measured particle velocities and erosion in water and air flows. Wear **263**(1), 330–338 (2007). https://doi.org/10.1016/j.wear.2006.12.048
4. J.F. Archard, Contact and rubbing of flat surfaces. J. Appl. Phys. **24**(8), 981–988 (1953). https://doi.org/10.1063/1.1721448

5. B. McLaury, J. Wang, S. Shirazi, J. Shadley, E. Rybicki, Solid particle erosion in long radius elbows and straight pipes, in *SPE Annual Technical Conference and Exhibition*, (San Antonio, 1997)
6. M.A. El-Emam, L. Zhou, E. Yasser, L. Bai, W. Shi, Computational methods of erosion wear in centrifugal pump: a state-of-the-art review. Arch. Comput. Methods Eng. **29**(6), 3789–3814 (2022). https://doi.org/10.1007/s11831-022-09714-x
7. C. Kang, Q. Cao, S. Teng, H. Liu, K. Ding, Wear characteristics of a centrifugal pump transporting solid–liquid mixture: an experimental and numerical study. Ain Shams Eng. J. **15**(1), 77 (2024). https://doi.org/10.1016/j.asej.2023.102277
8. H. Wang, Z. Tan, S. Kuang, A. Yu, Systematic investigation of centrifugal slurry pump under different operating condition by DDPM method. Powder Technol. **430**, 24 (2023). https://doi.org/10.1016/j.powtec.2023.119024
9. H. Wang, Z. Tan, S. Kuang, A. Yu, Numerical modeling and analysis of particle-fluid flow and wall erosion in centrifugal slurry pumps under different solid concentrations. Powder Technol. **410**, 861 (2022). https://doi.org/10.1016/j.powtec.2022.117861
10. K. Wang, H. Liu, L. Wang, P. Guo, Y. Wang, J. Yang, Effect of particle size on vortex structure and erosion behavior of semi-open centrifugal pump. Energy **293**, 576 (2024). https://doi.org/10.1016/j.energy.2024.130576
11. X. Wen et al., CFD-DEM investigation of vortex characteristics within the hydraulic jet pumps in the vertical hydraulic conveying process. Chem. Eng. J. **507**, 576 (2025). https://doi.org/10.1016/j.cej.2025.159576

Chapter 15
Applications to Mining

Mining is the process of extracting valuable minerals and other geological materials from the Earth's crust. These materials, which include metals like gold and copper, industrial minerals such as limestone, and energy resources like coal, are essential for modern infrastructure, technology and energy production. Mining operations can be conducted on the surface or underground, depending on the depth and distribution of the ore body. The extracted raw ore typically contains a mix of valuable minerals and waste rock, necessitating further processing to separate and concentrate the desired materials.

As shown in Fig. 15.1, mining processing often begins with comminution, a critical step that reduces the ore's particle size to liberate the target minerals for subsequent separation and refinement. Comminution is highly energy-intensive, accounting for up to 50% of the energy consumption of mining. However, this energy-intensive process is very important, as smaller particles increase the surface area for following chemical or physical separation methods like flotation or leaching. Comminution occurs in stages, each tailored to achieve specific size reductions while optimizing energy efficiency. The process begins with primary crushing, where large ore chunks which is often up to 1 m in diameter are broken down to about 10–20 cm using jaw or gyratory crushers. Secondary and tertiary crushing further reduce the particles to finer sizes of about 5–10 mm through cone or impact crushers. Crushed ore is then screened to classify particles by size, with oversize material recycled back into the crushers. For finer size reduction, the ore enters grinding mills, such as ball mills or semi-autogenous grinding (SAG) mills. SAG mills combine ore and grinding media such as steel balls to exploit both impact and abrasion forces, while ball mills rely primarily on grinding media to achieve sub-millimeter particle sizes. The product of grinding is fed to downstream processes like flotation, where minerals are separated based on surface chemistry.

Besides the crushers and mills, screens and chutes also play indispensable roles in ensuring efficient material handling in the mining process. Screens are critical for particle size classification, separating crushed ores into specific size fractions to

R. K. Agarwal and Y. Shao, *Theory and Application of Discrete Element Method to Simulation of Multiphase Flows*, https://doi.org/10.1007/978-3-032-11157-9_15

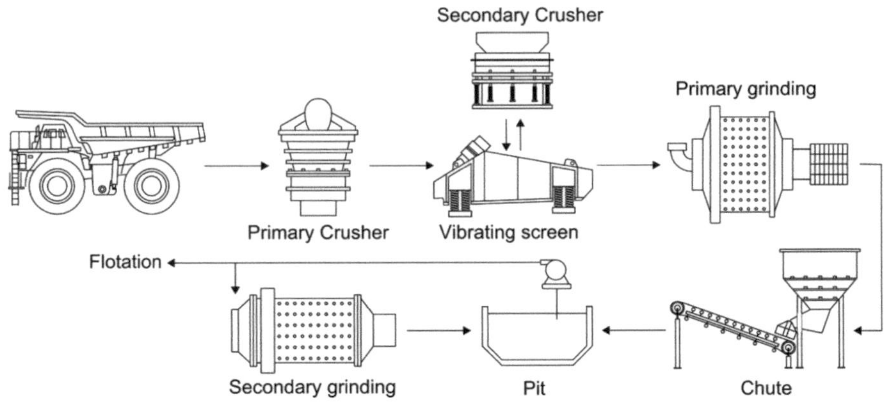

Fig. 15.1 Comminution flowchart [1]

meet downstream processing requirements, such as grinding, leaching or concentration, while also aiding in dewatering and removing fine impurities. Chutes, on the other hand, are essential for controlled material transfer between different stages of the mining process, such as from mills to storage bins, ensuring smooth, dust-free and energy-efficient ore flow. Together with crushers and mills, these components enhance operational efficiency, reduce equipment wear and minimize material blockages, directly impacting productivity, safety and cost-effectiveness in mineral processing plants.

15.1 Research Focus

DEM's ability to simulate the dynamic behavior of granular materials at the particle level has revolutionized the study of the equipment in mining process including crushers, mills, vibrating screens and chutes, enabling to analyze operational parameters, wear and energy efficiency in detail. Figure 15.2 displays the CFD-DEM research framework for mining.

In crushers, DEM has been extensively applied to study the dynamic interactions between particles and machine components, such as liners, mantles and concaves. Research focuses on modeling breakage mechanisms using approaches like particle replacement models (PRM) and bonded particle models (BPM). These models allow for the prediction of particle size distributions by simulating the fracture of particles under compressive and impact forces. The coupling of DEM with multibody dynamics has further enhanced the analysis of crusher performance by incorporating the dynamic response of moving parts, such as the eccentric motion in gyratory crushers. This integration helps optimize design parameters, including the geometry of crushing chambers and operational settings like rotational speed and closed-side

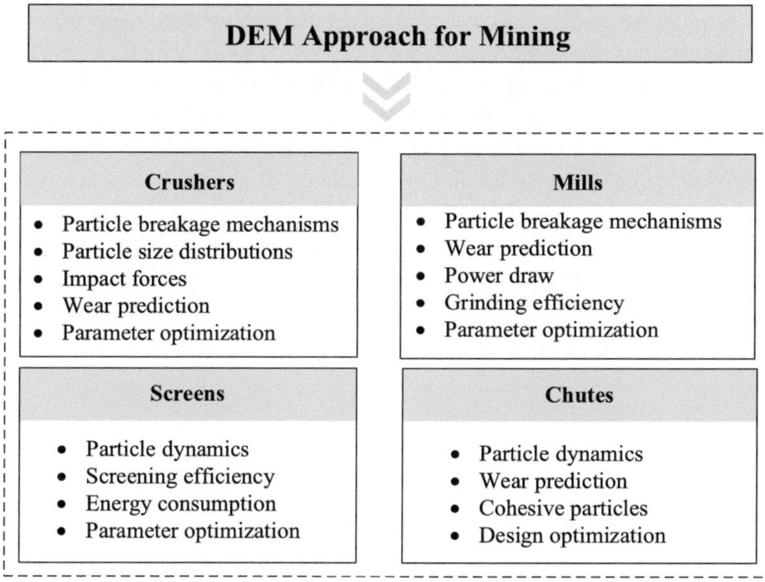

Fig. 15.2 CFD-DEM research framework for mining

setting. Despite these advancements, challenges remain in accurately calibrating fracture models to reflect real material behavior, particularly for heterogeneous ores.

The application of DEM in mills, such as ball mills and SAG mills, has provided valuable insights into liner wear, charge motion and particle breakage. Wear prediction models, such as the shear impact energy model and Archard's equation, are commonly integrated with DEM to assess the degradation of mill liners under different operating conditions. Simulations reveal how variables like mill speed, fill level and lifter design influence the distribution of impact and shear forces, which directly affect wear patterns and energy consumption. DEM has also been coupled with CFD and smoothed particle hydrodynamics to model the behavior of slurry within mills, offering a comprehensive understanding of grinding efficiency. The use of non-spherical particles in DEM simulations has improved the accuracy of predictions by better representing the irregular shapes of real ore particles. These advancements contribute to the development of strategies for prolonging liner life and optimizing grinding performance.

Screens which are critical for particle classification have been studied using DEM to evaluate screening efficiency and the effects of design and operational parameters. The method allows for the analysis of particle trajectories, stratification and passage through screen apertures under varying vibration amplitudes, frequencies and deck inclinations. Simulations have been employed to investigate the dynamic response of screen meshes, particularly in flip-flow screens, where flexibility plays a significant role in performance. Research efforts are also directed toward understanding wet screening conditions, where liquid bridges between particles introduce additional complexities. These studies aim to optimize screen designs for enhanced efficiency and reduced energy consumption.

In chutes, DEM has been instrumental in addressing challenges related to material flow, wear and blockages. Simulations analyze the influence of chute geometry, particle properties, and operational conditions on flow patterns and segregation. DEM–FEM coupling has enabled the prediction of structural wear and stress distribution, aiding in the design of more durable chute systems. Studies on cohesive materials, such as wet or fine particles, have provided insights into flow obstructions and methods to mitigate them. The method has also been applied to optimize transfer points in bulk material handling systems, ensuring smooth material flow and minimizing wear-induced downtime.

15.2 Example Applications: Crushers

As shown in Fig. 15.3, the gyratory crusher operates on the principle of compressive crushing through a gyrating mantle mounted on an eccentric shaft in a fixed concave bowl. As the mantle rotates eccentrically, it creates a cyclical motion that progressively narrows the crushing chamber, exerting compressive forces on the ore particles fed from the top. The ore is crushed by repeated compression between the mantle and concave, with the fragmented material discharging through the bottom due to gravity.

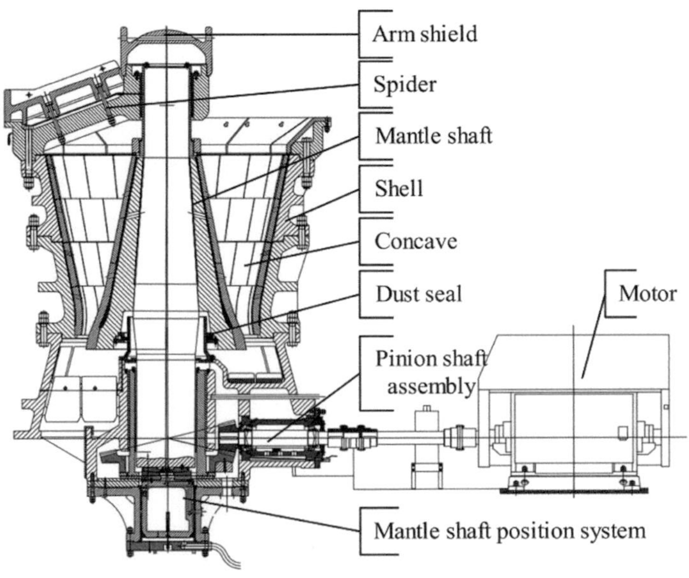

Fig. 15.3 Vertical cross-section of a gyratory crusher [2]

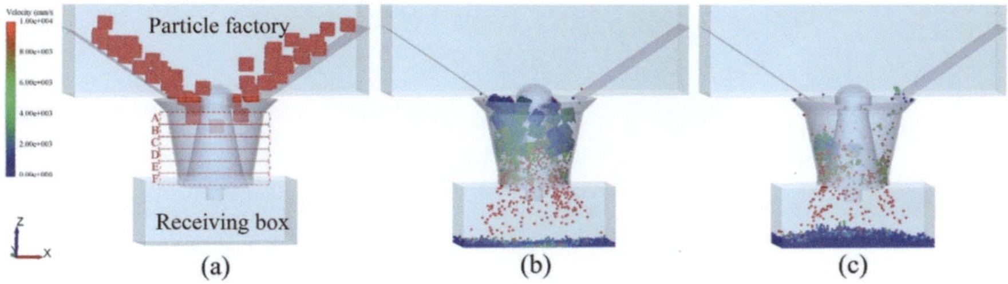

Fig. 15.4 The simulation of the working process of the gyratory crusher [2]: **a** $t = 0.66$ s; **b** $t = 8$ s; **c** $t = 16$ s

Chen et al. [2] focused on the simulation and optimization of gyratory crusher performance using the DEM and response surface methodology. As shown in Fig. 15.4, the study specifically analyzes the crushing process of iron ore, establishing a DEM model based on the BPM to simulate ore fragmentation. Key parameters such as the mantle shaft angle (MSA), concave angle (CA), eccentric angle (EA) and mantle shaft speed (MSS) are investigated for their impact on crusher performance, including productivity, crushing force and particle size distribution. The results reveal that MSA and the interaction between MSA and MSS significantly influence productivity, while MSA and CA have a strong effect on crushing force. The discharge granularity is most affected by MSA, MSS and their interactions. Multi-objective optimization, prioritizing productivity with crushing force as a secondary goal, yields optimal parameter values: MSA of 13°, CA of 14°, EA of 0.29° and MSS of 129.63 rpm. Additionally, the optimized parameters extend the effective crushing distance within the chamber, improving ore fragmentation uniformity.

The cone crusher operates on the principle of compressive crushing through an eccentrically gyrating mantle in a stationary concave bowl, as shown in Fig. 15.5. The mantle, mounted on a rotating eccentric shaft, performs a pendulum-like motion, creating a progressively narrowing gap that crushes ore particles fed from the top. As material moves downward by gravity, it undergoes repeated compression between the mantle and concave, with fragmentation achieved via interparticle breakage (layer-to-layer crushing) and direct contact with the liners. Different from gyratory crushers featuring a fixed concave with a gyrating mantle for high-capacity primary crushing, cone crushers use a rotating eccentric bushing and smaller, more versatile crushing chambers for secondary/tertiary crushing with finer output.

Delaney et al. [4] presented a DEM model for simulating non-spherical particle breakage and flow in an industrial-scale cone crusher. The study focuses on extending DEM capabilities to handle non-round particles, represented as super-quadrics, which are broken into non-round progeny based on compressive energy at contacts. A dynamic packing algorithm is employed to densely pack daughter fragments into the parent particle's volume, maintaining mass and volume conservation while resolving only coarser progeny in the simulation to manage computational costs. The simulations reveal that breakage predominantly occurs near the choke point where the mantle and concave are closest, with asymmetric feed leading to uneven wear on

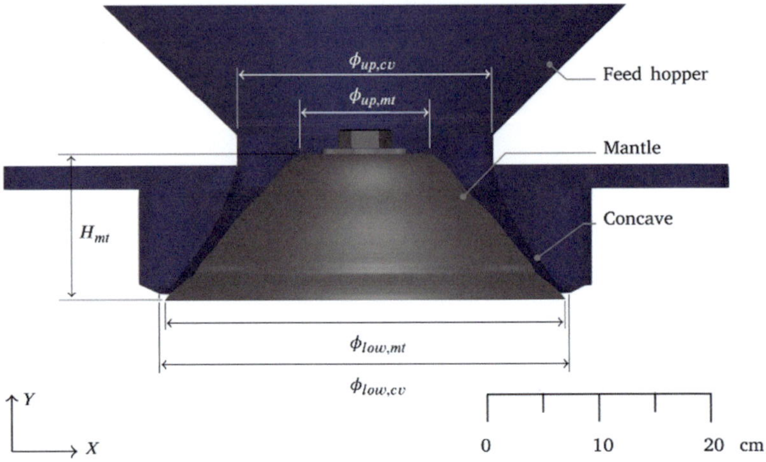

Fig. 15.5 Geometrical model of the cone crusher [3]

crusher liners, as shown in Fig. 15.6. The wear patterns show compressive damage concentrated in narrow vertical bands on both the mantle and concave, while abrasive wear is more broadly distributed. The study also demonstrates how particle forces and velocities vary cyclically with the mantle's motion, influencing breakage dynamics.

An impact crusher consists of a rotor equipped with hammers or blow bars that rotate at high speeds, striking the incoming feed materials, as shown in Fig. 15.7. When the materials enter the crusher, they are struck by the rotating hammers and thrown against impact plates or liners, causing them to fracture into smaller pieces. The repeated collisions between the particles, rotor, and impact plates further reduce the material size until it passes through the crusher's discharge opening. It is widely used due to its ability to handle hard and abrasive materials.

Chen et al. [5] conducted a numerical investigation of an industrial-scale impact crusher using the DEM integrated with the fast-cutting breakage model (FCBM). The

Fig. 15.6 Location of breakage events in the cone crusher over a three second interval [4]

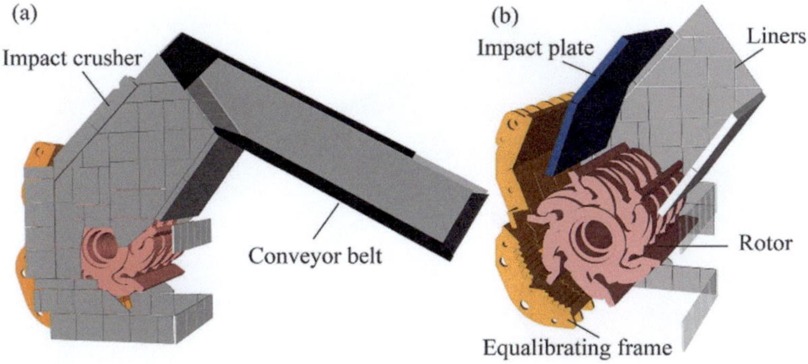

Fig. 15.7 **a** The impact crusher and its associated conveyor belt; **b** various components of the impact crusher [5]

study aims to simulate the breakage process of polyhedral particles in the crusher, as shown in Fig. 15.8. The FCBM, an improved breakage model, incorporates a breakage criterion based on fracture energy and a fast-cutting technique to generate polyhedral progeny particles, enabling efficient and accurate simulation of particle breakage. Key findings reveal several important aspects of crusher performance. The power consumption fluctuates periodically, with peaks exceeding 4000 kW when large particles collide with the rotor, followed by lower energy consumption as particles fragment further. This dynamic behavior underscores the importance of optimizing rotor speed to balance efficiency and wear. The wear primarily concentrates on the rotor and impact plates, with distinct spatial patterns. On the rotor, wear is highest at the hammerhead, increasing along the direction perpendicular to the rotational axis due to higher collision velocities at the outer edges. Meanwhile, the impact plate experiences more severe but sporadic wear on its upper half due to high-energy impacts from large rebounding particles, while the lower half shows uniform wear from accumulated particle collisions. The rotor speed significantly influences crusher performance. While increasing speed enhances breakage efficiency up to a critical point, further increases do not improve product fineness but instead raise power consumption and accelerate wear. Additionally, higher speeds cause more particles to rebound into the conveyor belt area, posing operational challenges. The feeding rate exhibits a linear relationship with power consumption, but within a certain range, it does not substantially affect product quality, suggesting that production can be scaled without compromising efficiency.

15.3 Example Applications: Mills

A tumbling ball mill is a traditional grinding machine which consists of a horizontally mounted rotating cylinder partially filled with grinding media (typically steel or ceramic balls) and the material to be ground. As the cylinder rotates, the balls are

Fig. 15.8 Schematic diagram of particle velocity distribution inside the impact crusher [5]

lifted and then cascade or drop onto the material, crushing it through impact and abrasion.

Li et al. [6] developed a DEM-based scale-up model to predict the performance of tumbling ball mills shown in Fig. 15.9 by analyzing damping energy distributions and their relationship with particle breakage. The mills of different sizes with 127–254 mm in diameter operating under various rotation speeds (34–100 rpm) and particle filling levels (0.4–1.0), using steel balls and limestone particles are simulated. The findings reveal that particle–ball (P–B) contacts are the primary mechanism for particle breakage, contributing the majority of effective damping energy despite being less frequent than particle–particle (P–P) contacts. The spatial distribution of damping energy shows that the highest energy impacts occur at the toe of the mill charge, where falling balls and particles collide most intensely. An optimal rotation speed exists where grinding rates peak-around 60 rpm in the tested setup. Beyond this speed, power draw continues to increase, but grinding efficiency declines because more energy is wasted in ball-ball collisions and particle centrifuging rather than productive breakage. Higher particle filling levels slightly reduce grinding rates by restricting ball movement but remain preferable for maximizing throughput. Larger mills demonstrate superior efficiency, as particles experience greater kinetic energy from longer free-fall distances, leading to proportional increases in grinding performance with mill size.

Unlike traditional ball mills that rely entirely on steel balls for grinding, a SAG mill combines the ore itself as the grinding medium (autogenous grinding) with a small addition of steel balls to enhance efficiency. The mill consists of a large rotating drum lined with wear-resistant materials, where the ore is fed and crushed by the impact of tumbling rocks and grinding media. SAG mills and are often used in the first stage of grinding before further refinement in ball mills.

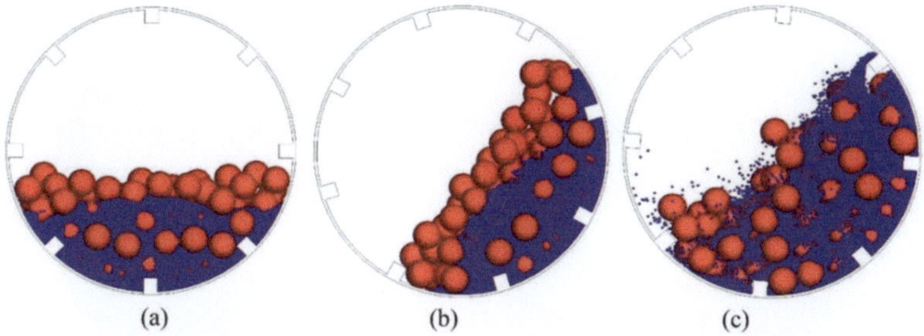

Fig. 15.9 Particle flow in a mill at different stages: **a** packed bed; **b** lifting stage; and **c** steady flow. Red: steel grinding balls; blue: limestone particles [6]

Zhu et al. [7] investigated the effect of lifter configuration on wet grinding performance and wear in an industrial-scale SAG mill shown in Fig. 15.10 using a GPU-powered CFD-DEM model coupled with the VOF method to account for slurry interactions. The research focuses on varying lifter height, face angle, number and shape to analyze their impact on mill dynamics, grinding efficiency and lifter wear. Key findings reveal that lifter height and number significantly influence grinding performance, with taller and more numerous lifters enhancing cataract load, breakage rates and energy utilization efficiency. In contrast, increasing the lifter face angle reduces power consumption by allowing particles to roll more easily, improving energy efficiency without substantially affecting grinding performance. The isosceles trapezoidal lifters outperform right-angled trapezoidal designs due to larger spacing between lifters, which enhances particle flow. Slurry presence reduces breakage rates, particularly for smaller particles, by damping collision energy. Wear analysis shows that the lifting and top faces of lifters experience the most severe wear, exacerbated by lower lifter hardness.

15.4 Example Applications: Screens

A vibrating screen generally consists of a sloped or horizontal screen deck mounted on a vibrating frame. An electric motor drives an eccentric shaft or vibrator mechanism, generating high-frequency vibrations that stratify and propel particles across the screen surface. Smaller particles pass through the apertures, while larger particles travel to the discharge end.

Zhao et al. [8] investigated the motion characteristics of particles on vibrating flip-flow screens using a coupled FEM-DEM approach. The research focuses on two main models: a single elastic screen plate for particle stratification analysis and a double-deck vibrating flip-flow screens for industrial-scale screening simulations. The single elastic screen plate model reveals that particle stratification is most effective at an excitation frequency of 70 rad/s, where distinct displacement differences between particle sizes (3–25 mm) are observed. Larger particles exhibit

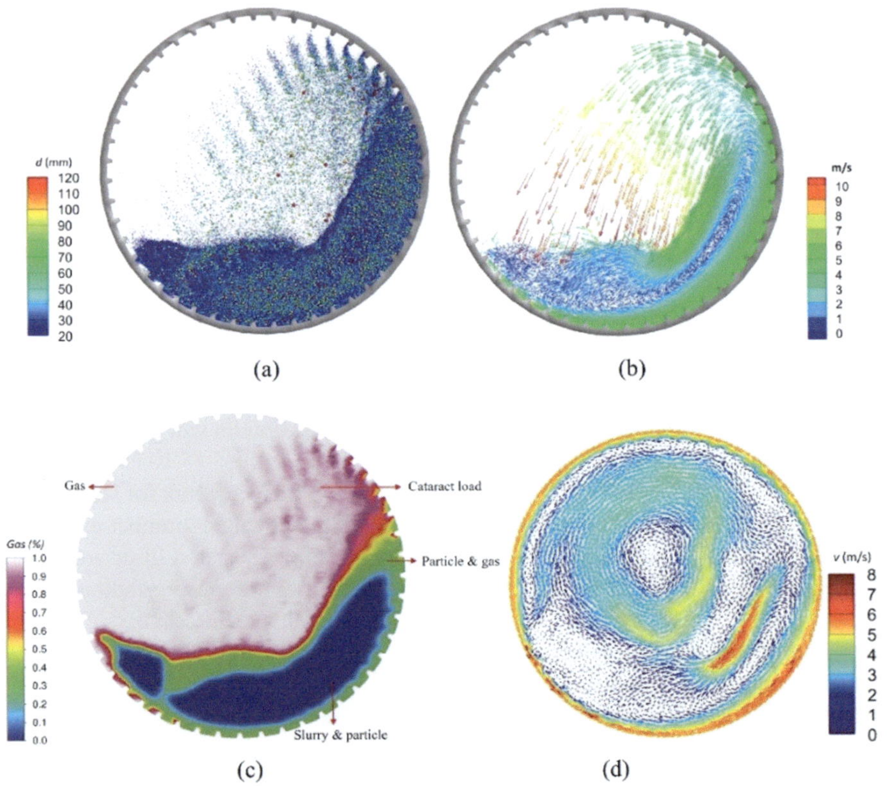

Fig. 15.10 Snapshots of the SAG mill behavior at $t = 50$ s: **a** particle flow pattern colored by particle diameter; **b** particle velocity vector; **c** gas volume fraction; and **d** fluid velocity vector [7]

higher vertical displacements due to their kinetic energy, while finer particles remain closer to the screen surface. The study identifies three motion phases: collision, separation and rise/fall, with particle velocity peaking during collisions and declining during stratification. The optimal frequency balances energy transfer and stratification, minimizing particle mixing. For the double-deck vibrating flip-flow screens shown in Fig. 15.11, the research categorizes particle motion into three regions: stable (translational motion), transitional (wave-like motion) and chaotic (irregular jumps) regions. The "filtering effect" of the particle layer dampens high-frequency screen vibrations, retaining only fundamental frequencies in the stable zone. Transitional and chaotic zones exhibit higher-frequency responses, aiding in particle loosening and screen self-cleaning. Simulations show that feed rates of 150–200 kg/s minimally affect particle velocity along the screen surface but reduce perpendicular velocity for finer particles due to layer thickness. Coarse particles of 16–25 mm maintain consistent velocities, while finer particles of 3–5 mm show greater variability. Overall, the double-deck vibrating flip-flow screens demonstrate efficient separation, with coarse particles discharged as oversize and fine particles penetrating the lower deck.

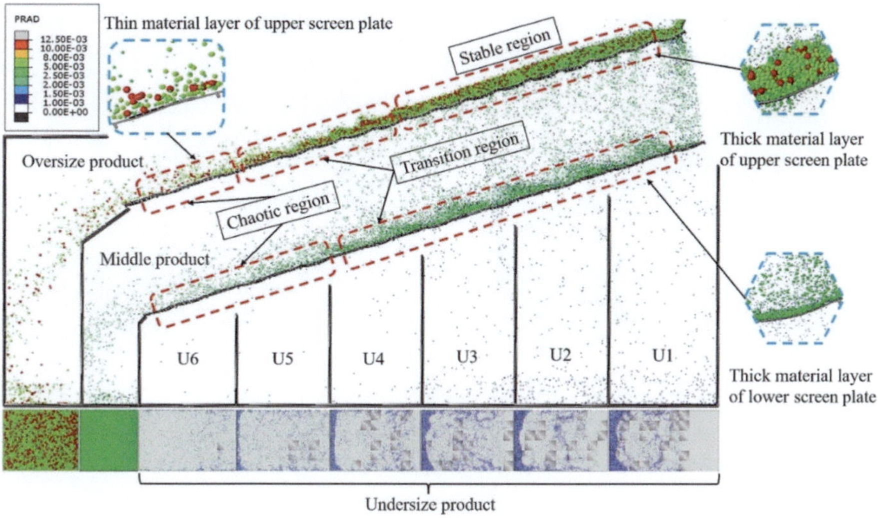

Fig. 15.11 Numerical simulation diagram of particle movement zone and particle screening process on the surface of the double vibrating flip-flow screen [8]

A roller screen is used primarily for coarse material separation and is less common than the vibrating screen in the mining process. It features a series of parallel rotating rollers which is often steel or rubber-coated spaced at defined intervals to sort materials by size. As the rollers turn, smaller particles fall through the gaps while larger materials are conveyed forward for further processing. Unlike vibrating screens, roller screens rely on mechanical rotation rather than vibration, making them ideal for handling sticky, wet or oversized materials with minimal clogging.

Yang et al. [9] investigated the screening process of an industrial-scale roller screen shown in Fig. 15.12 using the DEM approach to analyze the effects of various operational and material parameters on screening efficiency. The roller screen is simulated with both spherical and non-spherical particles to evaluate the impact of feed rate, rotational speed of rollers, inclination angle of the screen deck and cohesion forces between wet particles. The results reveal that screening efficiency decreases with increasing feed rate but remains above 0.80 when the feed rate is below 700 kg/s. Optimal performance is achieved with a rotational speed of 100 r/min and an inclination angle of 6°, which maximizes efficiency by balancing particle transport and stratification. Particle shape is found to have minimal influence on screening efficiency, though non-spherical particles exhibit slightly higher efficiency under certain conditions due to their longer residence time on the deck. However, their lower mobility could hinder stratification at very high feed rates. The study also demonstrates that cohesion forces between wet particles have negligible effects on screening performance, indicating the roller screen's suitability for processing sticky materials. Additionally, it is also found particles with smaller particles passing

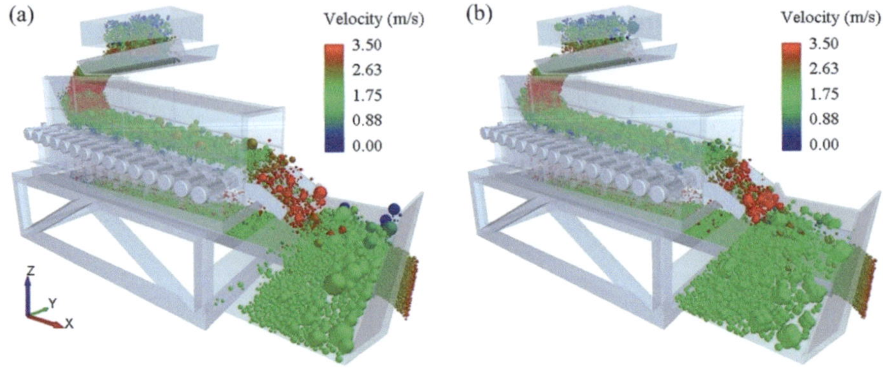

Fig. 15.12 Snapshots of the screening processes in DEM simulations with the feed rate of 600 kg/s, rotational speed of 80 r/min and inclination of 6° at around 10 s for spherical particles (**a**) and non-spherical particles (**b**) [9]

through the apertures earlier near the feed end, while near-aperture-sized particles require more time and are often screened near the discharge end.

15.5 Example Applications: Chutes

In mining, chutes are essential material handling structures designed to control and direct the flow of bulk solids between processing stages. Typically constructed from wear-resistant steel, rubber or ceramic-lined plates, chutes transfer materials from crushers, screens or storage bins to conveyors, trucks or processing equipment while minimizing dust, spillage and degradation. Key designs include transfer chutes, loading chutes and discharge chutes.

Nematollahi et al. [10] used the DEM approach to optimize the design of feed chutes in cone crushers shown in Fig. 15.13, aiming to improve performance such as power draw stability, throughput and product size distribution. The feed chute design is very important for achieving uniform feed distribution across the crusher's distribution plate, which is essential for minimizing segregation and enhancing crushing efficiency. Poor feed distribution leads to high power draw fluctuations, uneven wear on crusher liners and suboptimal product quality. Key findings reveal that transitioning from a cubic to a cylindrical feed chute design significantly improves performance. The cylindrical design, with reduced surface area and extended lengths above and below the feed chute plate, ensures more uniform feed distribution and vertical material flow into the crusher. DEM simulations demonstrate that the optimized feed chute reduces power draw fluctuations by 26%, increases throughput by 36%, and produces finer, more consistent product sizes. Additionally, the wear on crusher liners becomes more uniform, extending their operational life and reducing unplanned shutdowns.

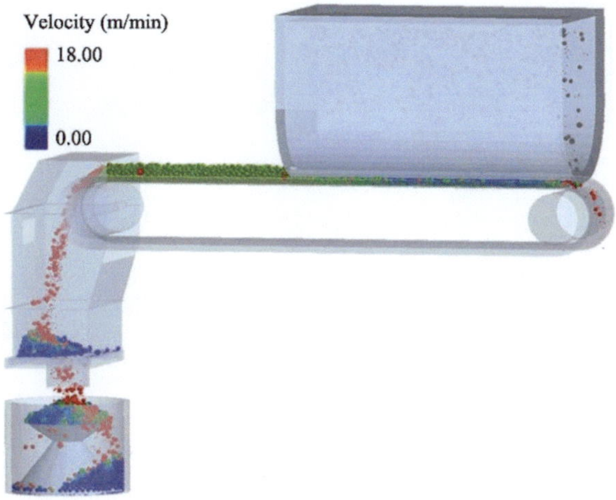

Fig. 15.13 Simulation of material flow in tertiary cone crusher circuit [10]

Jacobs-Capdeville et al. [11] focused on the analysis of energy dissipation during iron ore transfer through chutes with different designs shown in Fig. 15.14, employing a GPU-based DEM model. The study evaluates five chute head designs: traditional rock box, impact plate, and hood chutes, as well as modified micro-boxed versions of the impact plate and hood. The research aims to understand how energy dissipation mechanisms influence particle flow, wall wear, particle degradation and segregation during the transfer process. The rock box chute, while effective in flow redirection, suffers from high energy dissipation at the head section, leading to significant particle degradation and segregation. In contrast, the micro-boxed impact plate design reduces degradation by minimizing energy dissipation, and the traditional hood chute demonstrates superior mixing performance due to smoother particle flow redirection. The micro-boxed designs, incorporating small ledges to form particle beds, substantially decrease wall wear by reducing direct particle–wall collisions while maintaining flow control comparable to traditional chutes. The findings highlight a correlation between early-stage energy dissipation and increased segregation, suggesting that delaying energy loss in the transfer process improves particle mixing.

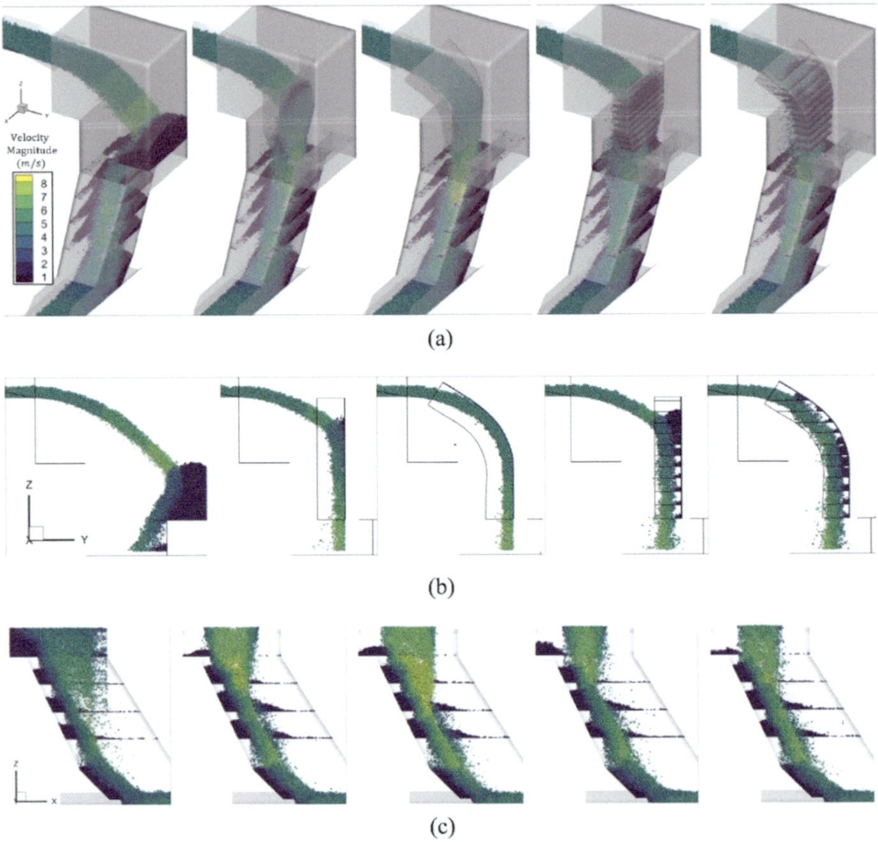

Fig. 15.14 Snapshots showing steady-state particle patterns colored with particle velocities: **a** transfer station overview, **b** head chute—viewed on a YZ slice and **c** lower sections—viewed on a XZ slice. From left to right: rock box, impact plate, hood, micro-boxed impact plate and micro-boxed hood head chute [11]

References

1. P. Toledo, M. Moncada, C. Ruiz, F. Betancourt, C.G. Rodríguez, C. Vicuña, *A Review of the Application of the Discrete Element Method in Comminution Circuits* (Elsevier, 2025). https://doi.org/10.1016/j.powtec.2025.121027
2. Z. Chen, G. Wang, D. Xue, Q. Bi, Simulation and optimization of gyratory crusher performance based on the discrete element method. Powder Technol. **376**, 93–103 (2020). https://doi.org/10.1016/j.powtec.2020.07.034
3. M. Moncada, P. Toledo, F. Betancourt, C.G. Rodríguez, Predictive modeling of crushing power in cone crushers with the discrete element method. Powder Technol. **447**, 178 (2024). https://doi.org/10.1016/j.powtec.2024.120178
4. G.W. Delaney, R.D. Morrison, M.D. Sinnott, S. Cummins, P.W. Cleary, DEM modelling of non-spherical particle breakage and flow in an industrial scale cone crusher. Miner. Eng. **74**, 112–122 (2015). https://doi.org/10.1016/j.mineng.2015.01.013

5. F. Chen et al., DEM simulation of an impact crusher using the fast-cutting breakage model. Powder Technol. **450**, 442 (2025). https://doi.org/10.1016/j.powtec.2024.120442
6. Y. Li, Y. You, D. Gou, A. Yu, R. Yang, A DEM based scale-up model for tumbling ball mills. Powder Technol. **409**, 854 (2022). https://doi.org/10.1016/j.powtec.2022.117854
7. Q. Zhu, Y. Feng, P. Witt, W. Bruckard, R. Yang, Investigation of the effect of lifter configuration on wet grinding and wear in an industrial-scale SAG mill with GPU-powered CFD-DEM modelling. Powder Technol. **458**, 54–92 (2025). https://doi.org/10.1016/j.powtec.2025.120959
8. G. Zhao, K. Pu, N. Xu, S. Gong, X. Wang, Simulation of particles motion on a double vibrating flip-flow screen surface based on FEM and DEM coupling. Powder Technol. **421**, 8422 (2023). https://doi.org/10.1016/j.powtec.2023.118422
9. X.D. Yang et al., DEM study of particles flow on an industrial-scale roller screen. Adv. Powder Technol. **31**(11), 4445–4456 (2020). https://doi.org/10.1016/j.apt.2020.09.020
10. E. Nematollahi, S. Zare, M. Maleki-Moghaddam, A. Ghasemi, F. Ghorbani, S. Banisi, DEM-based design of feed chute to improve performance of cone crushers. Miner. Eng. **168**, 68–78 (2021). https://doi.org/10.1016/j.mineng.2021.106927
11. P. Jacobs-Capdeville, S. Kuang, A. Yu, Unrevealing energy dissipation during iron ore transfer through chutes with different designs. Powder Technol. **435**, 13–19 (2024). https://doi.org/10.1016/j.powtec.2024.119446

Chapter 16
Applications to Pharmaceutical Manufacturing

Pharmaceuticals are medicinal products designed to diagnose, treat or prevent diseases, playing a crucial role in modern healthcare. These products include tablets, capsules, powders, granules and coated dosage forms, each requiring precise manufacturing processes to ensure efficacy, safety and consistency. The production of solid oral dosage forms involves multiple unit operations, such as milling, blending, granulation, tabletting and coating, each contributing to the final product's quality attributes.

Milling is a critical step in pharmaceutical manufacturing, where raw materials including active pharmaceutical ingredients (APIs) and excipients are reduced to a desired particle size. This process enhances dissolution rates, improves blend uniformity, and ensures consistent compaction during tabletting. Common milling techniques include hammer milling, ball milling, jet milling and conical scree milling, each suited for different material properties. For example, jet milling is preferred for heat-sensitive APIs due to its low-temperature operation, while hammer milling is efficient for brittle materials. The particle size distribution obtained from milling directly impacts downstream processes, such as flowability in blending and compressibility in tabletting.

After milling, powders must be uniformly mixed to ensure dose consistency. Blending involves combining APIs with excipients such as diluents, disintegrants and lubricants. Common blending equipment includes tumble blenders (V-blenders, bin blenders) and high-shear mixers. The blending process must achieve homogeneity while avoiding segregation, especially for low-dose APIs where even minor variations can lead to potency deviations. Parameters such as blender speed, fill level and mixing time are carefully controlled to optimize uniformity.

Granulation converts fine powders into larger, free-flowing granules to improve compressibility and prevent segregation. There are two primary methods: wet granulation and dry granulation. Wet granulation involves adding a binding liquid such as water, ethanol or polymer solutions to the powder blend, followed by mechanical agitation to form moist agglomerates. These wet granules are subsequently dried

R. K. Agarwal and Y. Shao, *Theory and Application of Discrete Element Method to Simulation of Multiphase Flows*, https://doi.org/10.1007/978-3-032-11157-9_16

and sized to achieve the desired particle characteristics. This method significantly improves particle bonding, resulting in tablets with greater mechanical strength and reduced friability. Dry granulation, preferred for moisture-sensitive compounds, employs mechanical compaction through roller compaction or slugging to densify powders without liquid binders. The resulting compacts are then milled into uniform granules. Granulation is particularly important for formulations with poor flow properties or high-dose APIs, where uniform distribution is critical.

Tabletting or compression is the process of forming granules or powders into tablets using a rotary tablet press. The powder blend is fed into dies, where upper and lower punches apply high pressure to form a compact. Key tablet properties, such as hardness, disintegration time and dissolution rate, depend on compression force, dwell time and formulation composition. Challenges such as capping, lamination or sticking can occur if process parameters are not optimized.

Tablet coating applies a thin polymer layer to improve stability, control drug release or mask taste. The most common approach, film coating, involves spraying polymer solutions onto tablets in rotating pans or fluidized beds, creating a protective yet permeable film. For targeted drug delivery, enteric coatings using pH-sensitive polymers ensure release occurs in the intestines rather than the stomach. Traditional sugar coating, though less common today, builds multiple sugar syrup layers primarily for aesthetic improvement. Precise control of coating parameters including spray rate, air temperature and pan rotation speed is critical to prevent defects like sticking, uneven coverage or surface imperfections. This process not only improves tablet stability and drug release profiles but also enhances patient acceptability through taste masking and color uniformity.

16.1 Research Focus

DEM has emerged as a powerful computational tool for simulating and optimizing various pharmaceutical manufacturing processes. In milling operations, DEM simulations provide valuable insights into particle breakage mechanisms by modeling the complex interactions between grinding media and powder particles. These simulations can predict energy distribution patterns in milling chambers, enabling engineers to optimize parameters such as rotational speed, grinding media size and filling level to achieve desired particle size distributions while minimizing energy consumption. The method's ability to track individual particle trajectories helps identify dead zones in milling equipment and predict wear patterns on grinding surfaces, contributing to more efficient mill design and operation.

In blending processes, DEM plays a crucial role in understanding powder mixing dynamics. The simulations can accurately predict blending uniformity by modeling particle–particle and particle-equipment interactions under various operating conditions. DEM helps identify optimal blender configurations, rotation speeds and fill levels by visualizing segregation patterns and mixing kinetics. This is particularly valuable for formulations containing cohesive powders or components with significant density differences, where conventional trial-and-error approaches would be time-consuming and resource-intensive. The method's ability to simulate different particle shapes and size distributions allows formulators to predict blending behavior for complex multi-component systems, significantly reducing development timelines.

For granulation processes, DEM simulations offer unique capabilities in understanding particle agglomeration mechanisms. In wet granulation, DEM coupled with CFD can model liquid binder distribution and its impact on granule growth. The simulations help optimize binder addition rates, mixing intensity and granulation time by predicting granule size evolution and porosity development. In dry granulation, DEM accurately simulates powder compaction behavior during roller compaction, predicting ribbon density variations and subsequent granule properties after milling. These simulations enable process engineers to design more efficient granulation processes with reduced material waste and improved product consistency.

In tabletting operations, DEM provides fundamental understanding of powder compression behavior at the particle level. The method can simulate die filling dynamics, predicting potential segregation issues and optimizing feed frame designs. During compression, DEM models reveal how particles rearrange, deform and bond under pressure, helping to predict tablet mechanical properties and identify potential defects such as capping or lamination. The simulations allow virtual testing of different formulation compositions, enabling formulators to assess the impact of excipient properties on tablet quality without extensive experimental work. DEM's ability to model elastic recovery post-compression is particularly valuable for predicting tablet ejection forces and potential sticking problems.

For coating processes, DEM simulations track tablet motion in coating equipment, providing critical insights into coating uniformity. The method can predict residence time distributions in different regions of the coating pan, helping to optimize pan speed, baffle design and spray patterns. DEM simulations also model droplet-tablet interactions, enabling better understanding of coating film formation and drying kinetics. This information is crucial for developing robust coating processes, especially for functional coatings where uniformity directly impacts drug release characteristics. The ability to simulate different tablet shapes and sizes allows for efficient scale-up from laboratory to production equipment.

Overall, the integration of DEM across these pharmaceutical manufacturing processes enables a more scientific approach to process development and optimization. By providing particle-level insights that are difficult or impossible to obtain experimentally, DEM reduces development costs, accelerates scale-up and improves product quality. As computational power continues to increase and models become more sophisticated, DEM's role in pharmaceutical manufacturing is expected to expand further, potentially enabling real-time process monitoring and control through digital twin technologies. The method's ability to bridge the gap between microscopic particle behavior and macroscopic process performance makes it an invaluable tool in modern pharmaceutical development and manufacturing.

16.2 Example Applications: Milling

Krok et al. [1] investigated the breakage behavior of bonded pharmaceutical granules during the ball milling process using a combination of density-based clustering and the DEM. The primary objective is to understand the fragmentation mechanisms of granules with varying shapes and sizes under different milling conditions, particularly focusing on the effect of rotation speed, as shown in Fig. 16.1. Three granule shapes including elongated, blocky and rounded shapes are modeled, each with three sizes, to study their breakage patterns. Key findings reveal that rotation speed significantly influences the milling efficiency and fragmentation outcomes. At lower speeds of 47 rpm, granules exhibited rolling motion with poor mixing and minimal breakage, while cascading motion of 143 rpm improve mixing but still limit fragmentation. The most effective regime is cataracting motion of 190 rpm, where intensive collisions between granules, balls and lifters lead to high shear rates and optimal breakage. Conversely, centrifugal motion of 238 rpm reduces fragmentation due to radial segregation and diminishes ball-granule interactions. Additionally, the DEM simulations demonstrate that granule size has a more pronounced impact on breakage force than shape, with larger granules requiring higher forces to fragment.

Deng et al. [2] applied the DEM approach to simulate the operation of a conical screen mill (comil) shown in Fig. 16.2, a device commonly used for granule milling

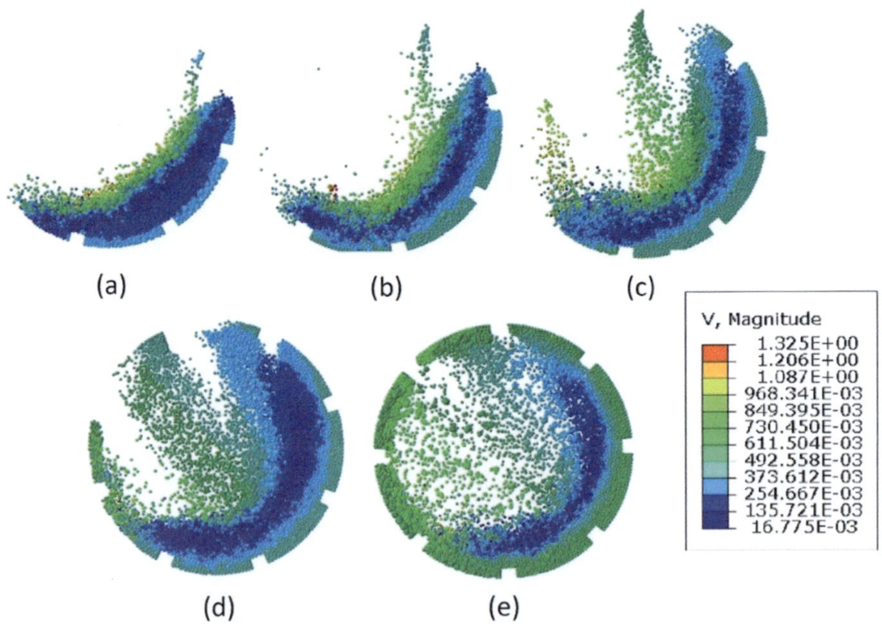

Fig. 16.1 The velocity distribution of flowing granules in the drum during ball milling with varying rotation speed: **a** 47 rpm; **b** 95 rpm; **c** 143 rpm; **d** 190 rpm and **e** 238 rpm [1]

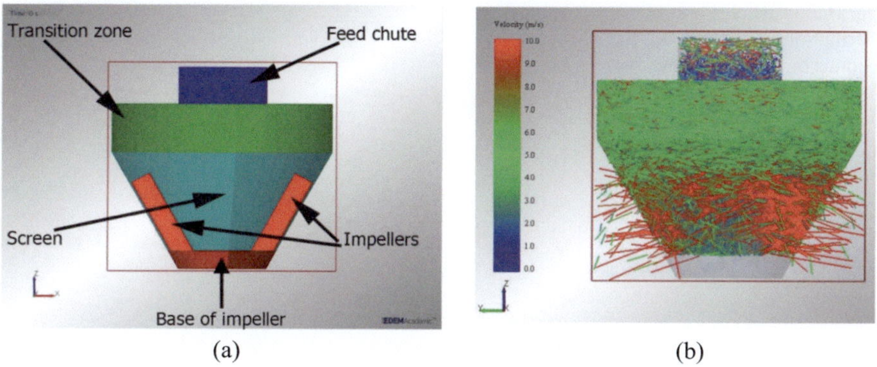

Fig. 16.2 **a** Schematic of the comil model; **b** particle trajectories inside the comil [2]

or cohesive powder delumping in pharmaceutical industry. The study investigates how various operating parameters, such as impeller speed, feed rate, screen hole size and open area, influence the mean residence time and residence time distribution of particles in the comil. Key findings show that mean residence time increases with higher impeller speeds. This behavior is attributed to the increased number of particles driven into the transition zone above the conical screen, where they remain longer due to reduced interaction with the screen holes. The simulation results also reveal that mean residence time decreases with higher feed rates, larger screen hole sizes and greater open areas, as these conditions facilitate faster particle exit from the comil. The average collision number and spatial distribution of particles are also examined, showing that higher impeller speeds lead to more collisions, particularly in the transition zone, which may enhance dry coating efficiency.

16.3 Example Applications: Blending

Behjani et al. [3] investigated the blending performance of pharmaceutical powder mixtures in a continuous twin-screw mixer shown in Fig. 16.3 using DEM. The primary focus is on assessing the formation and breakage of agglomerates of APIs, which are highly cohesive due to their fine particle size and strong interparticle forces. A sample-independent mixing index called the coefficient of blending performance (CBP) is introduced to quantify the homogeneity of the mixture by analyzing the contacts between API and excipient particles. Key findings reveal that the twin-screw mixer's geometry, particularly the inclusion of a kneading zone, significantly enhances blending efficiency by breaking API agglomerates. The kneading zone demonstrates the highest CBP values, indicating optimal mixing performance, while the secondary mixing zone shows poorer results due to re-agglomeration of APIs. The study also highlights the influence of operational parameters, such as rotational speed and particle surface energy, on blending quality. Lower rotational speeds of

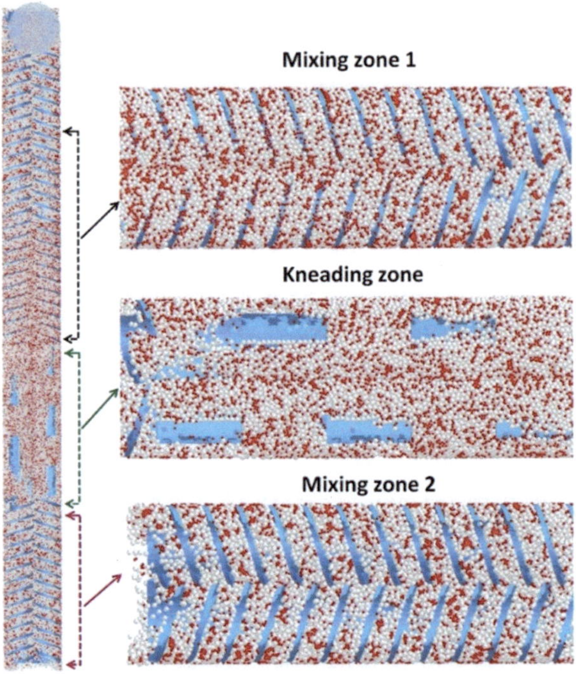

Fig. 16.3 Particles distribution in various sections of the twin screw mixer [3]

500 rpm in the primary mixing zone improve homogeneity, whereas higher speeds increase API-API contacts, leading to agglomeration. Additionally, higher surface energy values exacerbate cohesive behavior, further promoting agglomeration.

Sen et al. [4] investigated the mixing and segregation behavior of granular materials in a Bohle bin blender. The focus is on a multicomponent pharmaceutical blend consisting of coated theophylline granules (API) and excipients, aiming to understand the impact of critical process parameters including loading pattern, rotational speed and fill percentage on mixing dynamics. Key findings reveal that radial mixing (driven by top–bottom loading) is significantly faster than axial mixing (side-by-side loading) due to the symmetric geometry of the Bohle blender, which limits axial dispersion. The study also highlights the influence of operational parameters: higher rotational speeds of 20 rpm in the cascading regime enhance mixing rates for cohesive particles, while increased fill percentages of 40% reduce mixing efficiency due to restricted particle movement and insufficient headspace, as shown in Fig. 16.4.

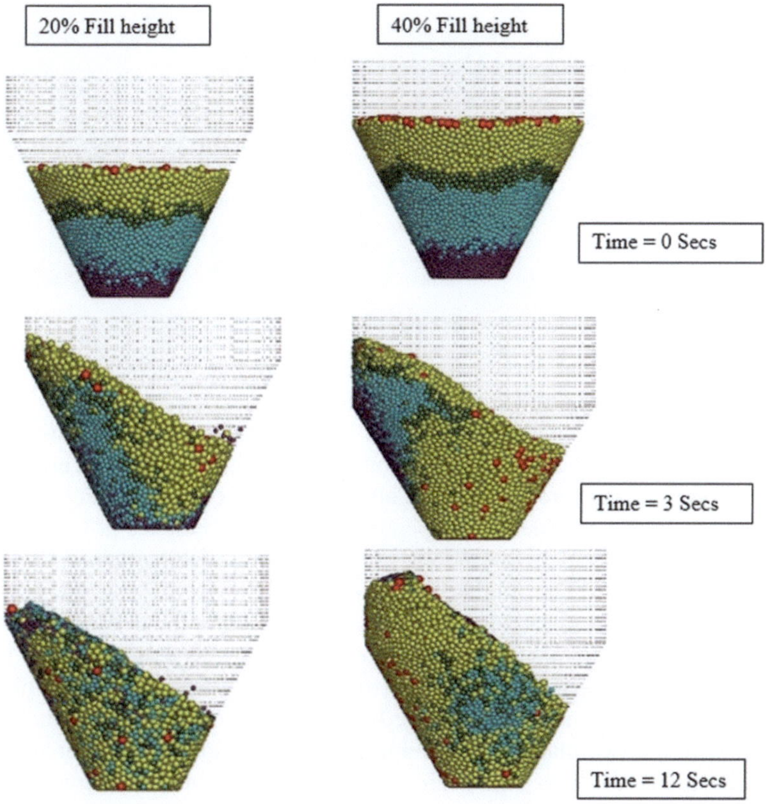

Fig. 16.4 Snapshots of mixing simulation of different fill percentage with 20 and 40% fill heights in a bin blender [4]

16.4 Example Applications: Granulation

Fries et al. [5] conducted the CFD-DEM modeling of fluidized bed spray granulators, with a particular emphasis on understanding particle and fluid dynamics at the individual particle level. The study compares two granulator configurations: a top-spray fluidized bed and a Wurster-coater, analyzing their performance in terms of particle wetting homogeneity and residence time distribution in the spray zone. Key findings reveal that the Wurster-coater shown in Fig. 16.5 exhibits a narrow residence time distribution, leading to more uniform particle wetting, whereas the top-spray granulator results in uneven wetting due to its irregular particle flow patterns. The research highlights the importance of process parameters such as jet velocity and the gap distance between the Wurster tube and the bottom plate. For instance, a stable circulating fluidization regime in the Wurster-coater is achieved when the jet velocity exceeds 100 m/s and the gap distance is kept below 15 times the particle diameter. Additionally, the results also show that the Wurster-coater produces narrower particle

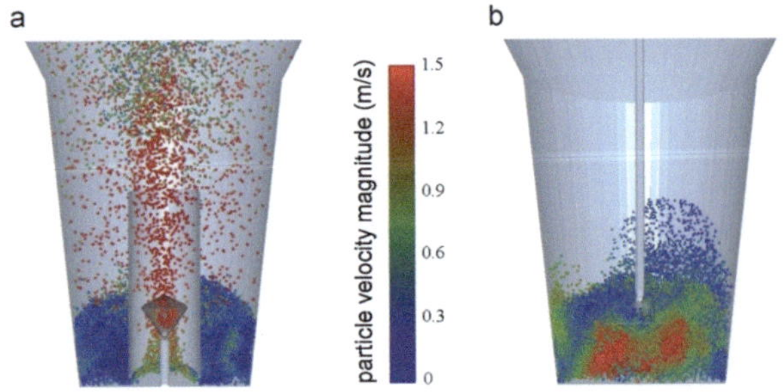

Fig. 16.5 Instantaneous particle positions and velocity distributions inside the Wurster-coater **a** and the top spray granulator, **b** at the simulation time $t = 1.4$ s [5]

size distributions compared to the top-spray configuration, underscoring the impact of homogeneous wetting on granulation efficiency.

Börner et al. [6] compared a novel two-blade impeller and a traditional three-blade impeller in high-shear wet granulation shown in Fig. 16.6, with an emphasis on mixing behavior, shear forces and granulation performance. Key findings reveal that the two-blade impeller achieves comparable mixing performance at lower rotation speeds, reducing thermal stress and energy consumption while maintaining robust granulation outcomes. The mixing investigations demonstrate that the two-blade impeller ensures uniform mixing across scales, adhering to the constant circumferential velocity scale-up rule, whereas the three-blade impeller exhibits longer mixing times post-scale-up due to its flow regime dependency. DEM simulations highlight that shear forces are primarily distributed in the bulk material rather than near the impeller, with the two-blade impeller applying slightly lower shear forces despite its reduced rotation speed. This design also minimizes kinetic energy disparities, reducing granule breakage rates compared to the three-blade impeller, which suffers from higher breakage due to elevated particle velocities.

16.5 Example Applications: Tabletting

Wu [7] investigated the die filling process in pharmaceutical tabletting using the DEM. The study focuses on the flow behavior of powders from a stationary shoe into a moving die, simulating conditions akin to rotary tablet presses, as shown in Fig. 16.7. The effects of particle shape, size, size distribution, initial powder bed height and filling speed on the average mass flow rate and critical filling speed, that is the maximum speed at which the die can be completely filled, are explored. The simulations employ 2D irregularly shaped particles, modeled as polygons, to better represent real-world pharmaceutical powders. The study reveals that poly-disperse systems, which mimic real powders with varied particle sizes, exhibit higher mass flow rates and critical filling speeds compared to mono-disperse systems. This is attributed to the enhanced flowability of poly-disperse powders due to better particle packing and reduced inter-particle friction. Additionally, particles capable of tessellation, such as squares and hexagons, result in lower filling rates and critical speeds, as

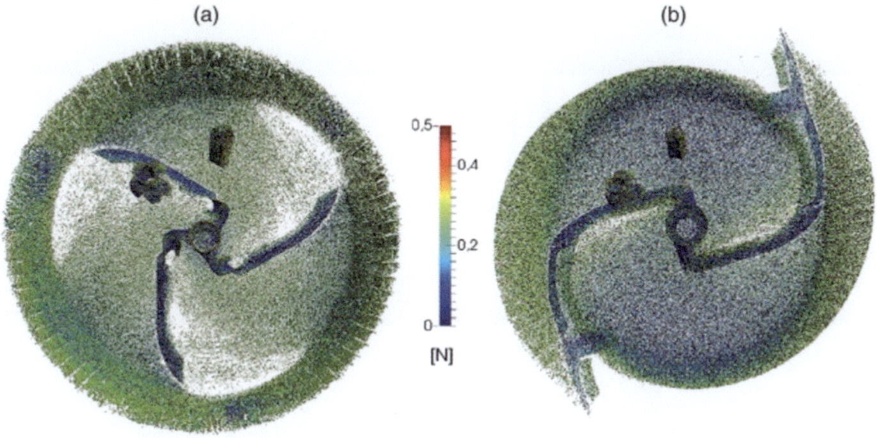

Fig. 16.6 Top view of instantaneous shear forces on particles in direct contact with the wall and the impeller for **a** the three-blade impeller at 90 min^{-1} and **b** the two-blade impeller at 30 min^{-1} at $t = 5$ s in high-shear wet granulation [6]

their ability to interlock impedes flow. In contrast, non-tessellating shapes like circles and pentagons facilitate smoother powder flow. Results also show that a higher initial bed height leads to increased filling rates and critical speeds, as the greater mass of powder exerts more pressure, promoting flow into the die. However, when the initial bed height is kept constant, varying the number of particles does not significantly alter the flow behavior, indicating that the initial bed height is a more critical factor than the absolute number of particles. The research further identifies two competing mechanisms during die filling: particle free-fall under gravity, which dominates at low speeds, and densification/jamming due to the die's horizontal movement, which becomes significant at higher speeds. The interplay of these mechanisms explains the observed peak in flow rate at the critical filling speed.

Hildebrandt et al. [8] explored the powder flow dynamics in a pharmaceutical tablet press force feeder using the DEM, as shown in Fig. 16.8. The study focuses on a lab-scale rotary tablet press equipped with a three-paddle-wheel force feeder, simulating real-world conditions to analyze powder behavior during die filling. The research aims to understand how powder flow in the force feeder influences tablet quality attributes, such as mass uniformity, API content consistency and particle segregation. The simulations capture the entire die filling process, from system initialization to high-speed tableting, and evaluate key metrics like mass flow rate, particle velocity, segregation and residence time distribution. The study reveals that the design of the force feeder, particularly the paddle wheel shapes and their arrangement, significantly impacts powder flow behavior. For instance, the reverse dosing wheel, with its cuboid-shaped blades, induces higher particle velocities and shear forces compared to the other paddle wheels, leading to increased particle attrition and segregation. This segregation results in non-uniform API distribution and tablet mass variability, which can compromise product quality. A critical finding is the

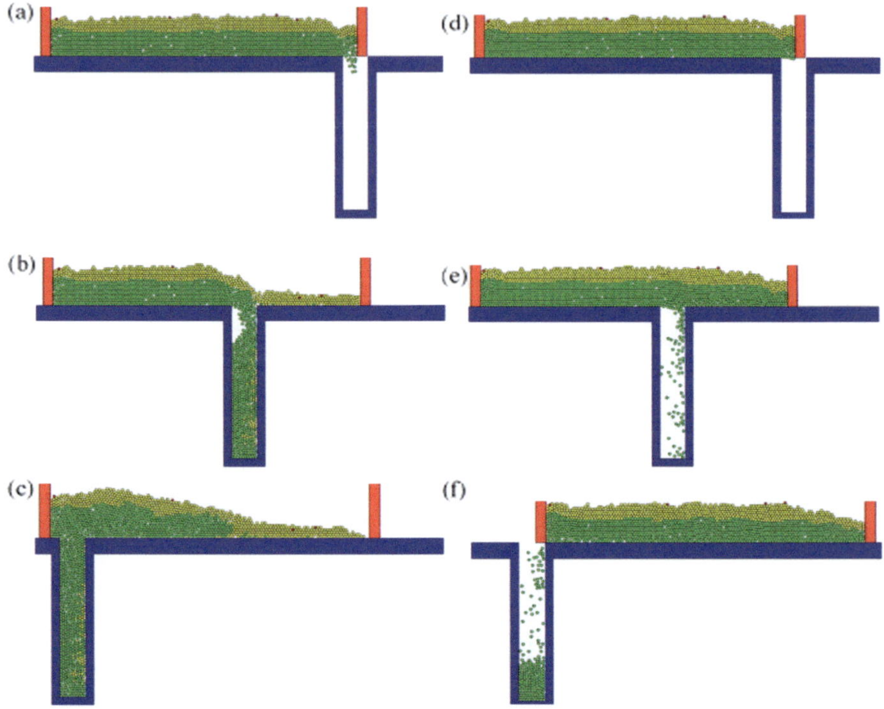

Fig. 16.7 Flow patterns during active die filling with a filling speed of 100 mm/s (**a–c**) and 400 mm/s (**d–f**), respectively [7]

identification of particle size segregation in the force feeder, where smaller particles tend to accumulate in specific zones due to mechanisms like sieving and trajectory segregation. This segregation leads to intra-die variability, with larger particles settling at the bottom of the die and smaller particles on top, potentially affecting tablet compaction and dissolution properties. The study also highlights the presence of a dead zone in the feeding hopper, where particles remain stagnant, contributing to powder consolidation and inconsistent flow.

16.6 Example Applications: Coating

Ketterhagen [9] conducted the DEM simulations on pharmaceutical tablet film coating to understand and optimize the coating uniformity, as shown in Fig. 16.9, which is critical for both cosmetic and functional coatings. The research highlights two main types of coating uniformity: inter-tablet uniformity, which refers to consistency between different tablets, and intra-tablet uniformity, which pertains to consistency across different sides of a single tablet. Inter-tablet uniformity is influenced

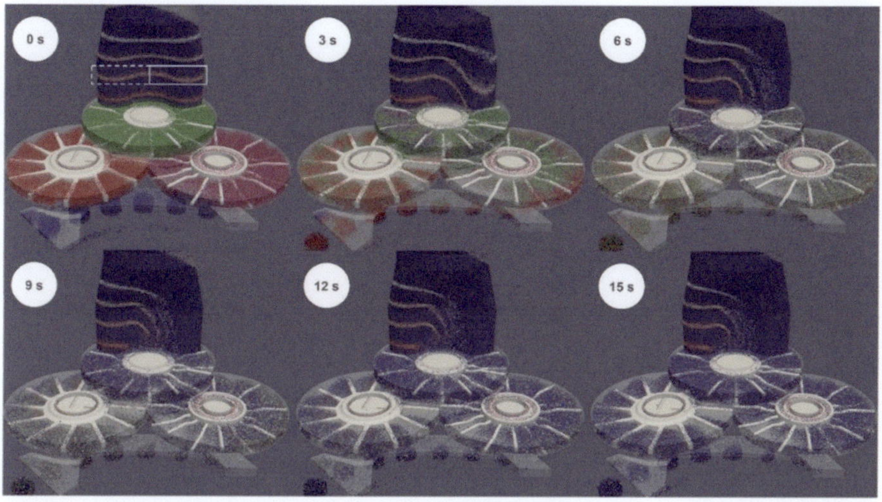

Fig. 16.8 Tracking of particles, colored by their initial position during high-speed tableting at intervals of 3 s in a pharmaceutical tablet press force feeder [8]

by tablet mixing dynamics and the distribution of tablets in the spray zone, while intra-tablet uniformity depends on the orientation of tablets as they pass through the spray zone. Key factors affecting these uniformities include tablet properties, process parameters and equipment geometry. Results show that non-spherical shapes, such as almond or shield tablets, tend to exhibit a preferred orientation in the spray zone, leading to uneven coating distribution across different sides of the tablet. In contrast, spherical or near-spherical shapes, like standard round concave (SRC) tablets, show more uniform orientations and better intra-tablet coating consistency. Process parameters such as pan speed and pan loading also impact coating uniformity, though their effects differ for inter- and intra-tablet variability. Higher pan speeds generally improve inter-tablet uniformity by reducing variability in residence and circulation times, while lower pan loadings achieve similar benefits by minimizing the likelihood of tablets bypassing the spray zone. However, these parameters have a relatively minor effect on intra-tablet uniformity compared to tablet shape.

Toschkoff et al. [10] presented a comprehensive study on the implementation and comparison of three different spray models for DEM simulations of particle coating processes in the pharmaceutical industry. The study focuses on drum coating shown in Fig. 16.10 to predict coating mass distribution, which is vital for achieving intra- and inter-tablet uniformity. The three methods investigated include an enhanced spray zone approach, a discrete drop method, and a ray-tracing method applied post-simulation. Key findings reveal that all three methods can produce comparable results, but their performance depends heavily on parameter choices. The spray zone approach, while straightforward, is sensitive to the grid spacing used for top detection, which can significantly affect the accuracy of the results. The discrete drop method, which models spray droplets as DEM particles, offers greater flexibility and

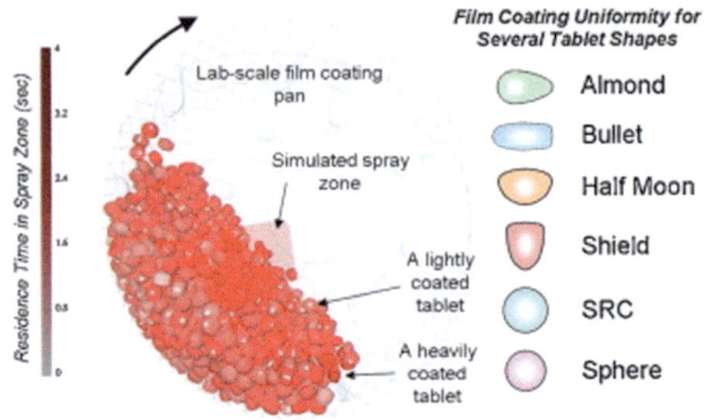

Fig. 16.9 DEM investigation of various pharmaceutical tablet shapes in a film coating pan [9]

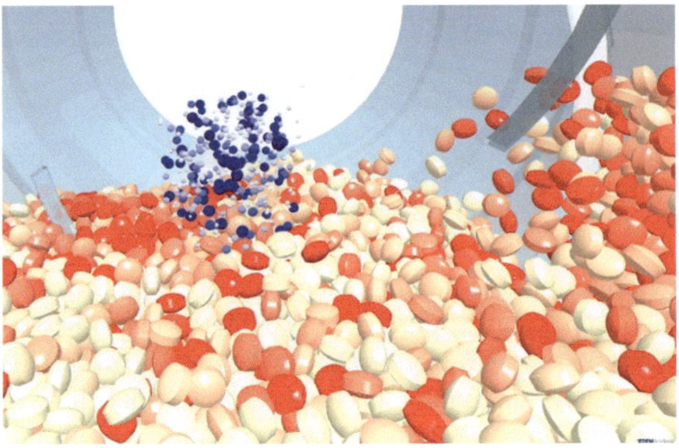

Fig. 16.10 DEM simulation of a tablet coating process, including the coating spray. The spray droplets are colored blue, with darker shades indicating bigger droplets. The tablets are colored according to coating mass, from white (no coating mass) to red (high coating mass) [10]

reliability, particularly when using a sufficient number of drops to ensure statistical accuracy. The ray-tracing method, though less accurate due to its reliance on post-processing and larger time steps, is faster and more flexible, making it suitable for preliminary analyses. It also emphasizes the importance of parameter selection, such as the number of drops or rays, spray interval and area uniformity, in achieving reliable simulations. Overall, the discrete drop method is identified as the most versatile, while the ray-tracing method provides a quick and efficient alternative for certain applications.

References

1. A. Krok, P. Peciar, K. Coffey, K. Bryan, S. Lenihan, A combination of density-based clustering method and DEM to numerically investigate the breakage of bonded pharmaceutical granules in the ball milling process. Particuology **58**, 153–168 (2021). https://doi.org/10.1016/j.partic.2021.03.008

2. X. Deng, J. Scicolone, X. Han, R.N. Davé, Discrete element method simulation of a conical screen mill: a continuous dry coating device. Chem. Eng. Sci. **125**, 58–74 (2015). https://doi.org/10.1016/j.ces.2014.08.051
3. M.A. Behjani, Y.G. Motlagh, A.E. Bayly, A. Hassanpour, Assessment of blending performance of pharmaceutical powder mixtures in a continuous mixer using discrete element method (DEM). Powder Technol. **366**, 73–81 (2020). https://doi.org/10.1016/j.powtec.2019.10.102
4. K. Sen, N. Velez, C. Anderson, J.K. Drennen, A.S. Zidan, B. Chaudhuri, Multicomponent granular mixing in a Bohle bin Blender-Experiments and simulation. Int. J. Pharm. **578**, 765–785 (2020). https://doi.org/10.1016/j.ijpharm.2020.119131
5. L. Fries, S. Antonyuk, S. Heinrich, S. Palzer, DEM-CFD modeling of a fluidized bed spray granulator. Chem. Eng. Sci. **66**(11), 2340–2355 (2011). https://doi.org/10.1016/j.ces.2011.02.038
6. M. Börner, M. Michaelis, E. Siegmann, C. Radeke, U. Schmidt, Impact of impeller design on high-shear wet granulation. Powder Technol. **295**, 261–271 (2016). https://doi.org/10.1016/j.powtec.2016.03.023
7. C.Y. Wu, DEM simulations of die filling during pharmaceutical tabletting. Particuology **6**(6), 412–418 (2008). https://doi.org/10.1016/j.partic.2008.07.008
8. C. Hildebrandt, S.R. Gopireddy, R. Scherließ, N.A. Urbanetz, Investigation of powder flow within a pharmaceutical tablet press force feeder: a DEM approach. Powder Technol. **345**, 616–632 (2019). https://doi.org/10.1016/j.powtec.2019.01.040
9. W.R. Ketterhagen, Modeling the motion and orientation of various pharmaceutical tablet shapes in a film coating pan using DEM. Int. J. Pharm. **409**(1–2), 137–149 (2011). https://doi.org/10.1016/j.ijpharm.2011.02.045
10. G. Toschkoff et al., Spray models for discrete element simulations of particle coating processes. Chem. Eng. Sci. **101**, 603–614 (2013). https://doi.org/10.1016/j.ces.2013.06.051

Chapter 17
Applications to Battery Technology

17.1 Electrode Calendering in Lithium-Ion Batteries

Lithium-ion batteries (LIBs) are critical for energy storage in applications ranging from portable electronics to electric vehicles. A key step in LIB electrode manufacturing is calendering, where electrodes are compressed to optimize porosity, particle connectivity and adhesion, thereby improving energy density and electrochemical performance. As shown in Fig. 17.1, the calendering process involves compressing the electrode coating, which is composed of active materials (AM), conductive additives and binders, between two rotating rollers to achieve a uniform thickness and controlled porosity. By adjusting parameters like roller pressure, temperature, and speed, calendering ensures the electrode's structural integrity and enhances its electrochemical properties.

17.1.1 Research Focus

The DEM has emerged as a valuable computational tool for studying the calendering process of battery electrodes, offering unique insights into the microscale phenomena that govern electrode performance. The method provides a virtual laboratory where researchers can observe particle rearrangement, deformation and fracture mechanisms that are otherwise difficult to measure experimentally.

The typical DEM workflow for calendering simulations begins with the generation of a virtual electrode layer that replicates the actual composition and particle size distribution. This digital representation is then subjected to compressive forces that mimic the calendering process, using appropriate contact models to simulate particle interactions. Advanced simulations may incorporate temperature effects or binder migration to increase realism. Following the simulation, detailed analysis reveals key metrics such as porosity distribution, particle connectivity and stress–strain response.

R. K. Agarwal and Y. Shao, *Theory and Application of Discrete Element Method to Simulation of Multiphase Flows*, https://doi.org/10.1007/978-3-032-11157-9_17

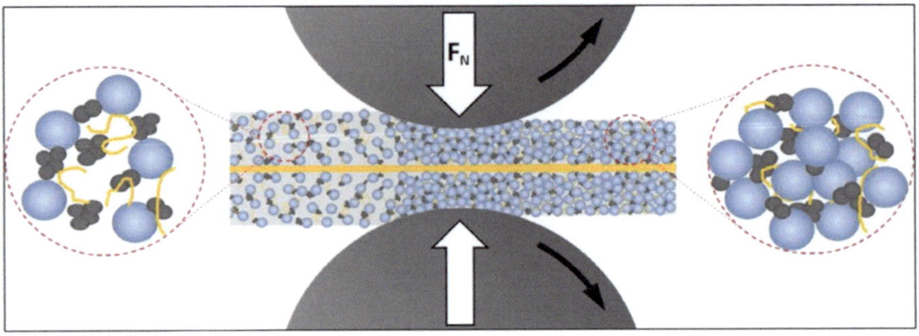

Fig. 17.1 Schematic illustration of the changes in electrode microstructure during the calendering process. The blue spheres represent active materials, the black spheres represent carbon and the yellow stripes represent binders [1]

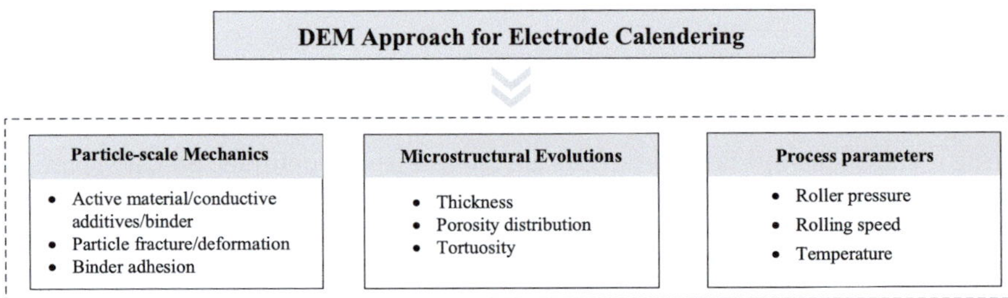

Fig. 17.2 DEM approach for electrode calendaring in lithium-ion batteries

The detailed research focus of DEM simulations on electrode calendaring is shown in Fig. 17.2.

(1) Particle-Scale Mechanics

One of the research focuses of DEM simulations for electrode calendering is the study of particle-scale mechanics. Active material particles, which are often brittle, may fracture under high pressure, while conductive additives like carbon may deform elastically. The binder, which provides mechanical integrity to the electrode, plays a particularly important role as it forms flexible bridges between particles. DEM models must accurately represent these binder interactions, typically using cohesive contact models or viscoelastic spring-damper systems, to capture their adhesive and energy-dissipating properties.

(2) Microstructural Evolutions

As the electrode undergoes compression, particles experience complex interactions including sliding, rolling and compaction. The DEM simulations capture how these processes lead to microstructural evolutions including thickness, porosity and tortuosity, which are critical for electrode performance. Thickness reduction improves electronic conductivity by enhancing particle contact but must balance active material loading for optimal energy density. Porosity decreases as voids collapse, enhancing electronic pathways at the cost of increased ionic resistance, necessitating a trade-off between energy density and electrolyte accessibility. Simultaneously, tortuosity rises as pore networks become more convoluted. DEM enables precise optimization of these interdependent parameters to guide electrode design.

(3) Influence of Process Parameters

Another major focus of DEM calendering studies is the influence of process parameters such as roller pressure, rolling speed and temperature on final electrode properties. Higher pressures generally increase electrode density but may also lead to excessive porosity reduction or particle damage. Rolling speed affects the homogeneity of compression, with faster speeds potentially causing uneven densification due to insufficient time for particle rearrangement. Temperature is particularly important for thermoplastic binders, which become softer and more adhesive when heated, altering the entire compaction dynamics. DEM simulations enable systematic investigation of these parameters, helping identify optimal processing conditions that balance electrode density, porosity and mechanical integrity.

(4) Electrochemical Performance

Beyond mechanical considerations, DEM simulations provide valuable insights into how calendering affects electrochemical performance. The formation of conductive networks is crucial for electron transport, and DEM can track whether conductive additives form percolating pathways throughout the electrode. Similarly, the evolution of pore structure directly impacts ion transport, with excessive compression potentially leading to high tortuosity and limited electrolyte infiltration. By quantifying these microstructural features, DEM helps establish processing-structure-performance relationships that guide electrode design. Furthermore, DEM can predict manufacturing defects such as particle agglomeration, density gradients or cracks that may form during calendering, enabling proactive process optimization.

17.1.2 Example Applications

Ge et al. [2] explored impact of calendering conditions on lithium-ion battery electrode microstructures using a combination of X-ray tomography and DEM simulations. The primary goal is to understand how compression during calendering affects electrode porosity, mechanical properties and transport characteristics, which are critical for battery performance. An experimentally obtained uncalendered cathode structure, composed of NMC ($LiNi_{1/3}Mn_{1/3}Co_{1/3}O_2$) particles, carbon black and PVDF binder is used as the initial structure for DEM simulations, as shown in Fig. 17.3. Results show that higher compression pressures lead to increased particle contact density and reduced pore volume, improving electrode density but potentially restricting ion transport. Under compression, force chains develop, indicating inhomogeneous stress distribution, as shown in Fig. 17.4. Fabric tensor analysis reveals increased contact anisotropy, with preferential alignment parallel to the current collector. Calendering increases tortuosity, potentially limiting ion diffusion but improving electronic conductivity. A binder phase generation algorithm demonstrates that carbon-binder domain distribution significantly influences tortuosity, highlighting the need for balanced electrode design.

While the earlier study established DEM as a viable tool for simulating electrode compression, it did not explore the direct link between microstructural modifications and electrochemical outcomes. Ge et al. [3] further conducted a research based on the prior work [2] which combined X-ray computed tomography and DEM to study porosity and tortuosity changes during calendaring, as shown in Fig. 17.5. The work extends this framework by incorporating finite element-based electrochemical modeling, enabling a comprehensive evaluation of how DEM-generated structures perform under realistic battery operating conditions. This advancement provides a more holistic understanding of the process-microstructure-performance relationship in LIB electrodes. The results demonstrate that calendering induces distinct

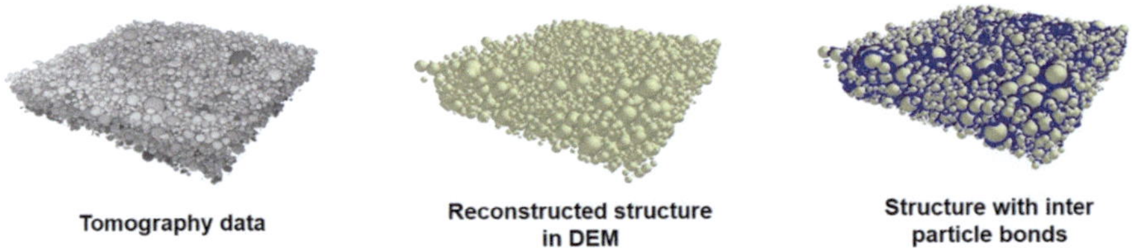

Fig. 17.3 Electrode structure generation used in DEM: tomography data processing [2]

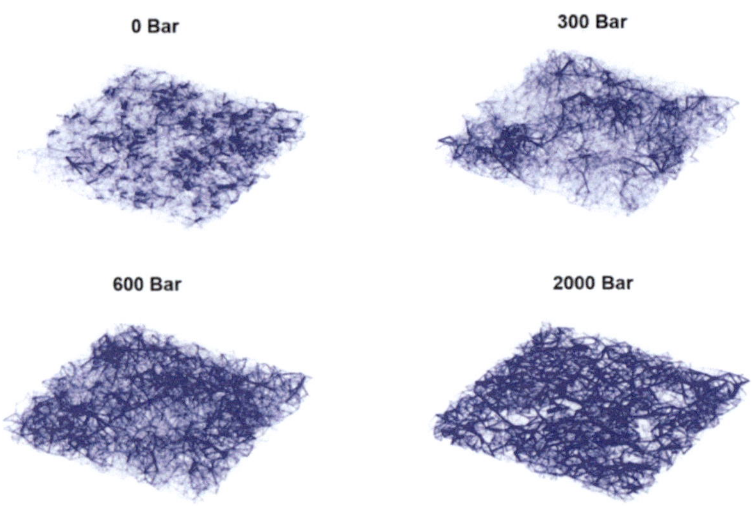

Fig. 17.4 Contact force network under different compression pressures [2]

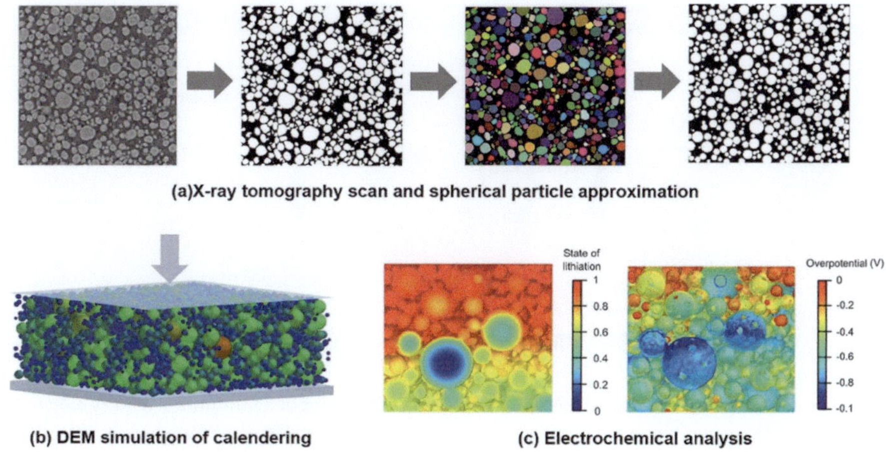

Fig. 17.5 DEM and electrochemical modeling of lithium ion cathode structures characterized by X-ray computed tomography [3]

microstructural changes, including reduced particle sphericity and increased alignment parallel to the current collector. While these modifications decrease overall porosity and increase tortuosity, which might be expected to hinder performance, the study reveals a counterintuitive improvement in rate capability. This enhancement is attributed to the reduction in electrode thickness, which shortens lithium-ion diffusion paths and outweighs the negative effects of increased tortuosity. The findings provide crucial insights into the complex trade-offs between porosity, thickness and electrochemical performance in battery electrode design.

Lundkvist et al. [4] also conducted a comprehensive DEM analysis of the mechanical behavior of positive electrode active layers of lithium-ion battery, with particular focus on the calendering process and subsequent mechanical characterization, as shown in Fig. 17.6. The research addresses the pressure-sensitive nature of electrode layers, which exhibit different responses in compression versus tension which is a critical factor in battery performance and durability. The DEM model incorporates two distinct contact formulations: an elastic–plastic Hertzian model for interactions between active material particles and a specialized binder contact model that accounts for both normal and tangential forces. The binder model treats the binder domain as a flat cylinder between particles, capable of elastic-perfectly plastic deformation. These formulations enable the simulation of the entire electrode lifecycle, from initial particle packing to calendering and mechanical testing. A representative volume element (RVE) approach with periodic boundary conditions is used to balance computational efficiency with accurate representation of the electrode microstructure. Results reveal distinct mechanical responses in compression versus tension, with compressive stiffness increasing by 40–60% due to growing particle contacts, while tensile stiffness remains stable as it is governed by the binder phase. During calendering, binder properties significantly influence outcomes: elastic binders show

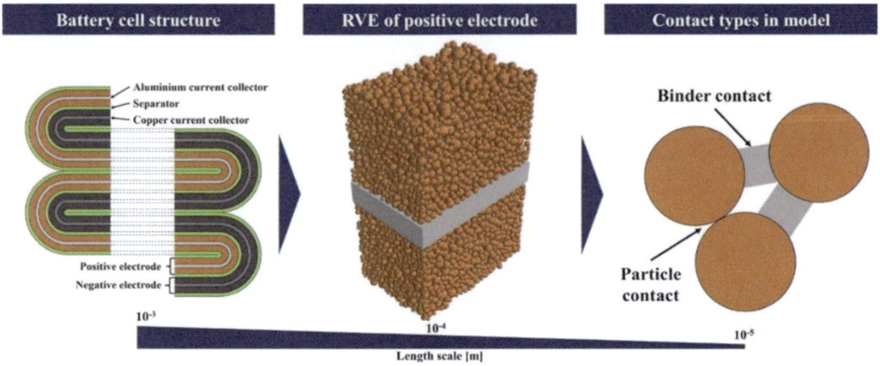

Fig. 17.6 A discrete element analysis of the mechanical behavior of a lithium-ion battery electrode active layer [4]

120% higher pressures when increasing the binder radius ratio, while plastic binders reduce pressures by 25–30% and substantially decrease elastic springback from 73 to 40%. Microstructural analysis shows the coordination number rising from 4.2 to 5.8 under compression, with binder contacts bearing 85–90% of tensile loads versus 60–70% for particle contacts in compression.

17.2 Particle Structure in All-Solid-State Batteries

All-solid-state batteries (ASSBs) represent a transformative advancement in energy storage technology, replacing the flammable liquid electrolytes of conventional lithium-ion batteries with solid electrolytes. This fundamental architectural shift eliminates safety risks associated with thermal runaway while enabling higher energy densities through the potential use of lithium metal anodes.

17.2.1 Research Focus

The solid-state nature of the ASSBs introduces new challenges, particularly regarding interfacial contact between rigid electrode and electrolyte particles, mechanical stability during cycling and manufacturing complexities. The DEM has emerged as a powerful computational tool to address these challenges by providing particle-scale insights into the microstructural evolution and interfacial behavior of ASSBs. As shown in Fig. 17.7, key research focus of application of DEM to ASSBs are described as follows:

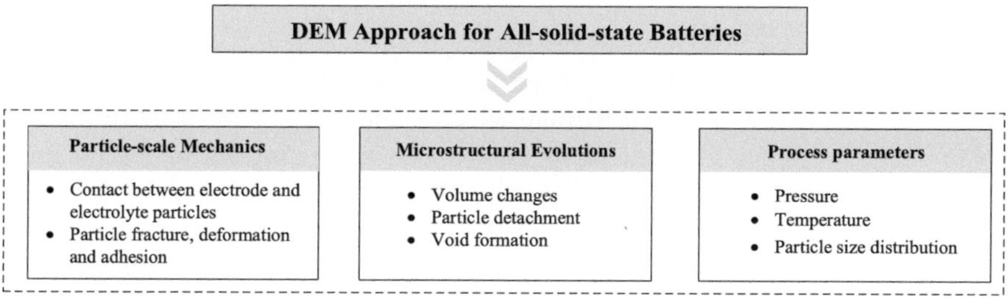

Fig. 17.7 DEM approach for particle structure in ASSBs

(1) Particle-scale mechanics

One of the most critical challenges in ASSBs is achieving and maintaining optimal solid–solid contact between electrode and electrolyte particles. Unlike liquid electrolytes, which can permeate porous electrodes and ensure uniform ionic transport, solid electrolytes rely on direct physical contact with AM particles. By modeling the interactions between particles such as deformation, fracture and adhesion, DEM helps predict the true contact area under different conditions.

(2) Microstructural evolutions

Another key application of DEM lies in studying the microstructural evolution of ASSBs during electrochemical cycling. Many high-capacity electrode materials, such as silicon or lithium metal, undergo significant volume changes during charge and discharge. These repeated expansions and contractions can lead to particle detachment, void formation and even fracture in the solid electrolyte. DEM simulations track these dynamic changes, capturing how microstructural rearrangements affect ionic and electronic conduction pathways over time. By incorporating electrochemical expansion models, DEM can predict stress accumulation at interfaces and identify failure mechanisms that lead to capacity fade. This knowledge is instrumental in designing more resilient electrode architectures, such as composite structures with buffering agents or engineered porosity to accommodate volume changes.

(3) Manufacturing processes

Manufacturing processes for ASSBs also benefit significantly from DEM modeling. The production of dense, defect-free solid electrolyte layers and well-integrated electrode–electrolyte interfaces requires precise control over powder compaction, sintering and lamination. DEM simulations can replicate these processes virtually, allowing researchers to optimize parameters such as pressure, temperature and particle size distribution. For instance, DEM has been used to study the cold pressing of sulfide solid electrolytes, revealing how particle rearrangement and fracture influence final density and ionic conductivity. Similarly, simulations of roll-to-roll processing help design scalable fabrication methods that ensure uniform interfacial contact without inducing mechanical damage.

(4) Multiphysics behaviors

To capture the complex multiphysics behaviors of ASSBs, advanced DEM frameworks now incorporate coupled thermal, electrochemical and mechanical models. These integrated simulations can predict how localized heating during fast charging affects interfacial reactions or how thermal expansion mismatches induce delamination. Multiscale approaches further enhance DEM's predictive power by linking particle-scale interactions with macroscopic battery performance. For example, DEM-FEM hybrid models analyze how cell-level stack pressure distributions influence local contact mechanics, while DEM-MD couplings provide atomic-scale insights into interfacial ion transport.

17.2.2 Example Applications

Lee et al. [5] employed a multiphysics modeling approach to investigate how initial pressure affects the mechanical and electrochemical performance of ASSBs. By integrating the DEM and FEM, the research bridges particle-level mechanical behavior with electrode-level electrochemical responses, as shown in Fig. 17.8. The DEM simulates the compaction of solid electrolyte and electrode mixtures, providing critical data on mechanical properties such as Young's modulus and volume fraction. These parameters are then incorporated into the FEM model, which evaluates electrochemical performance metrics like voltage and capacity, alongside mechanical responses such as stress and strain. A key finding of the study is the significant impact of pressure on electrochemical performance. Higher initial pressures enhance geometric properties, including surface area and volume fraction, which in turn improve electronic and ionic conductivity. However, the research also identifies a trade-off: while increased compression ratios boost capacity, they simultaneously elevate mechanical stress, raising the risk of electrode failure. This underscores the importance of balancing performance improvements with mechanical durability in battery design. The study further reveals that different battery components exhibit varying sensitivities to compression. The cathode demonstrates the greatest performance improvement under higher compression but also experiences the highest stress levels, making it a critical focus for optimization. In contrast, the solid electrolyte shows minimal sensitivity to compression changes, suggesting its role is less influential in pressure-dependent performance adjustments. Additionally, the research highlights the importance of external pressure, showing that even modest pressures can significantly enhance ionic conductivity and reduce interfacial resistance, though excessive pressure may lead to mechanical degradation.

Frankenberg et al. [6] investigated the production of composite cathodes for ASSBs using a high-intensity mixer, as shown in Fig. 17.9, focusing on the influence of stressing conditions on the electrochemical performance of the batteries using DEM. The composite cathode consists of submicron-sized $LiFePO_4$ (LFP) as the AM, carbon black (CB) for electronic conductivity, and Li_3InCl_6 (LIC) as the solid

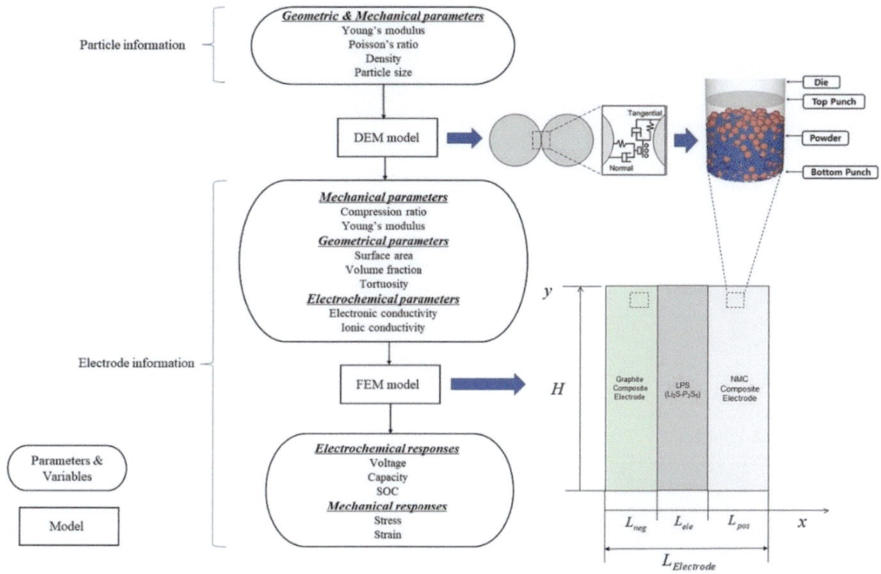

Fig. 17.8 Overall simulation flowchart and schematics of DEM and FEM for an ASSB [5]

electrolyte. These materials are mixed in a high-intensity mixer to form heteroaggregates, aiming to improve interfacial contact and reduce resistances. The mixer's rotational speed and mixing time are varied to study their effects on the composite's properties. The simulations reveal that higher rotational speeds increase both the stress intensity and frequency of particle collisions. Two primary stress mechanisms are identified: compression and shearing in the gap between the rotor and mixing drum, and impact stresses at the rotor blade. Experimental results show that higher rotational speeds and longer mixing times improve the discharge capacity of the ASSBs. This is attributed to better mixing and comminution of LIC agglomerates, and enhanced interfacial contact between the components. A novel force-scaling method is also introduced to handle coarse-grained particles in DEM simulations, ensuring accurate stress calculations while reducing computational time.

So et al. [7] conducted the DEM simulations of cold-pressing processes for ASSB cathodes. The focus is on comparing two fabrication methods: a conventional particle mixture and a novel solid electrolyte (SE) coating approach, as shown in Fig. 17.10. The goal is to elucidate the effects of these methods on ionic and electronic transport, mechanical damage and overall battery performance. The study simulates ASSB cathodes composed of AM particles and a sulfide-based SE. The DEM simulations model the cold-pressing process under varying mold pressures. The SE coating method involves attaching SE particles to the surface of active material aggregates, while the particle mixture method disperses SE particles randomly among AM particles. The simulations account for elastic and plastic deformation, particle aggregation, and the formation and breakage of fusion bonds between particles. Results

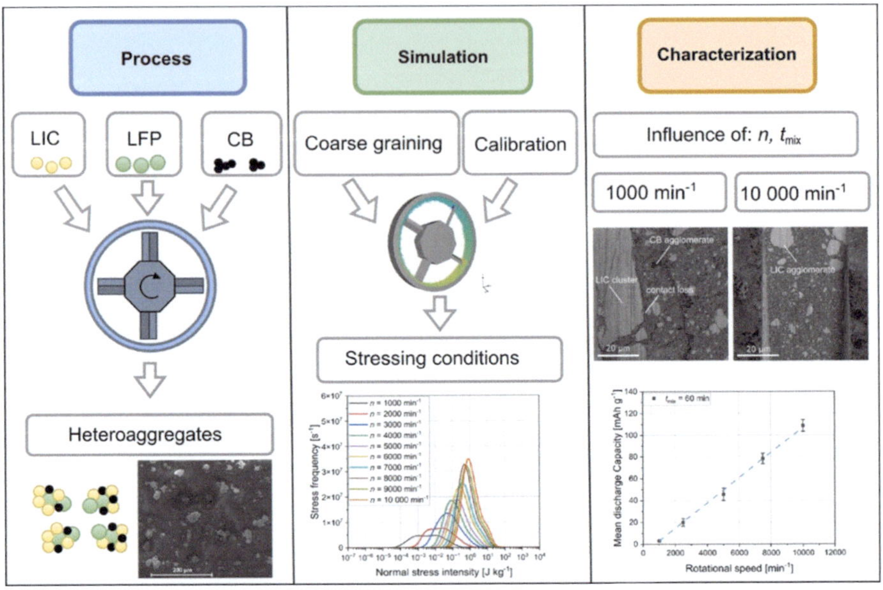

Fig. 17.9 Production of all-solid-state battery composite cathodes in a high-intensity mixer [6]

show that the SE coating method significantly improves ionic transport by creating a continuous percolation network, reducing tortuosity compared to the particle mixture method. At low mold pressures, the SE coating disrupts electronic pathways, but this effect diminishes at higher pressures of above 200 MPa due to SE delamination at AM contact points. The particle mixture method requires higher mold pressures of 360 MPa to achieve comparable SE percolation, whereas the coating method achieves over 90% percolation at just 25 MPa. Both methods exhibit similar AM damage profiles, with damage concentrated near the current collector and SE layer boundaries. The coating method does not provide significant protection against AM damage, as the SE's ductility does not mitigate stress concentrations during pressing. The SE coating undergoes delamination at AM contact points during pressing, but new SE particles reattach to AM surfaces, increasing the overall AM-SE contact area. This dynamic process ensures sustained ionic pathways even under high pressures. The SE coating method yields lower tortuosity for ionic transport, enhancing effective conductivity. In contrast, the particle mixture method performs better for electronic transport at lower pressures but converges with the coating method at higher pressures.

So et al. [8] presented a detailed simulation framework using the DEM to investigate the fragmentation of silicon (Si) AM in ASSBs, focusing on the interplay between volume expansion, diffusion limitations and mechanical degradation during charge–discharge cycles, as shown in Fig. 17.11. The research addresses a critical challenge in ASSBs: Si anodes, despite their high theoretical capacity, suffer from severe volume expansion of up to 300% during lithiation, leading to particle

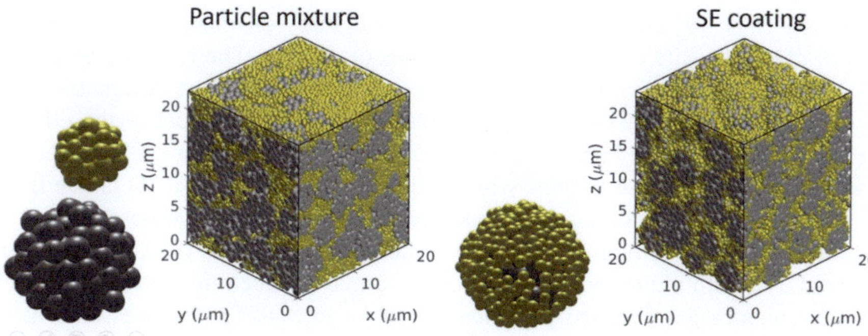

Fig. 17.10 Conventional particle mixture method and novel solid electrolyte coating method [7]

fragmentation, contact loss and rapid capacity decay. By integrating a multi-step DEM framework, which consists of particle initialization, cold pressing and intercalation simulations, the study provides a novel approach to quantify Si particle damage through fusion bond breakage and introduces a damage parameter to assess mechanical degradation. Key findings reveal that higher pressures reduce porosity and enhance AM-SE contact area but exhibit minimal impact on Si fragmentation. In contrast, charge–discharge rates (C-rates) significantly affect Li concentration gradients within Si particles. At high C-rates, uneven Li distribution induces localized stress concentrations, exacerbating particle fragmentation, while low C-rates mitigate damage by allowing more homogeneous diffusion. Additionally, increasing the AM fraction amplifies electrode stiffness but aggravates fragmentation due to reduced deformability.

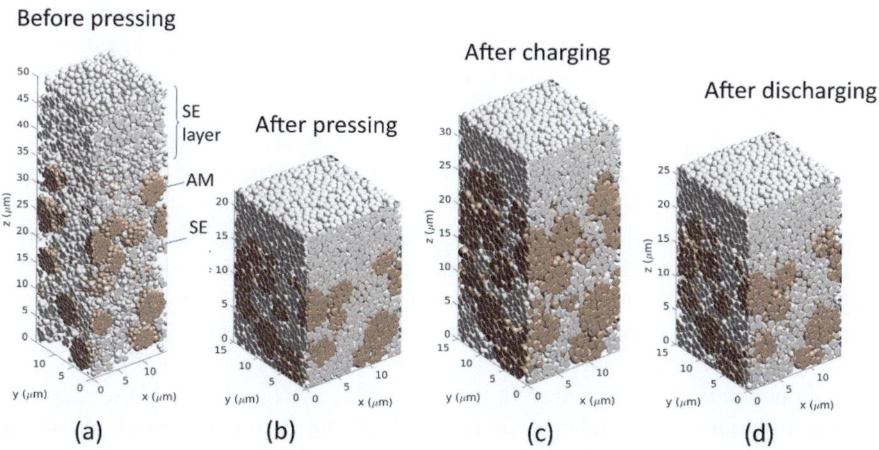

Fig. 17.11 The simulation steps conducted in simulation of the all-solid-state battery: **a** initialization, **b** cold-pressing, and intercalation after, **c** charging and **d** discharging [8]

Huang et al. [9] conducted a comprehensive study on modeling heat conduction in 3D composite cathode microstructures of ASSBs. The research focuses on understanding how microstructural parameters such as porosity, volumetric ratios of cathode AM and SE, particle size ratios, and tortuosity influence the effective thermal conductivity (ETC) of the cathode. The study employs a DEM to generate realistic cathode microstructures, which are then analyzed using a 3D lattice Boltzmann method as heat conduction model. Key findings reveal that porosity and the volumetric fraction of high-conductivity materials like $LiCoO_2$ (LCO) significantly impact ETC, with higher porosity reducing thermal conductivity and a higher LCO fraction enhancing it. Tortuosity, particularly of the pore and LCO phases, also plays a critical role, where increased tortuosity of pores improves ETC by reducing thermal resistance pathways, while higher tortuosity of LCO diminishes it. Additionally, smaller particle sizes are found to enhance thermal conductivity by providing more connected pathways for heat transfer.

17.3 Battery Recycling

Battery recycling is a critical process in addressing the environmental and economic challenges posed by spent batteries. The recycling process begins with discharging and dismantling the batteries for safety, followed by mechanical crushing to break them down into a mixture of plastic films, metal foils and electrode materials. The next step involves separating these components, where pneumatic separation plays a key role due to its cost-effectiveness and efficiency. This method leverages differences in particle density and aerodynamics, using controlled airflow to sort materials. Lighter plastics are carried upward while heavier metals settle downward. However, the irregular shapes of crushed battery particles present challenges for traditional separation techniques, necessitating advanced modeling tools for optimization.

DEM has emerged as a powerful computational approach to enhance pneumatic separation in battery recycling. Dianyu et al. [10] conducted a detailed numerical investigation and optimization of the vertical pneumatic separation process for recycling crushed lithium iron phosphate battery mixtures using a coupled CFD-DEM approach. The research focuses on the separation of non-spherical, film-like particles, including plastic films, anode copper foils and cathode aluminum foils, which are common components in spent LFP batteries. The primary objective of the study is to optimize critical operational parameters such as airflow velocity, separation zone height and initial particle velocity to maximize separation efficiency and recovery rates, thereby advancing battery recycling technologies. As shown in Fig. 17.12, the vertical pneumatic separator used in the study is a key component of the recycling process. It features a feed inlet where the crushed battery mixture enters, an airflow inlet that injects compressed air horizontally to fluidize the particles, and a vertical separation zone where the actual sorting occurs. Within this zone, lighter particles like plastic films are carried upward by the airflow, while heavier particles such as anode copper and cathode aluminum foils descend due to gravity. The separator also

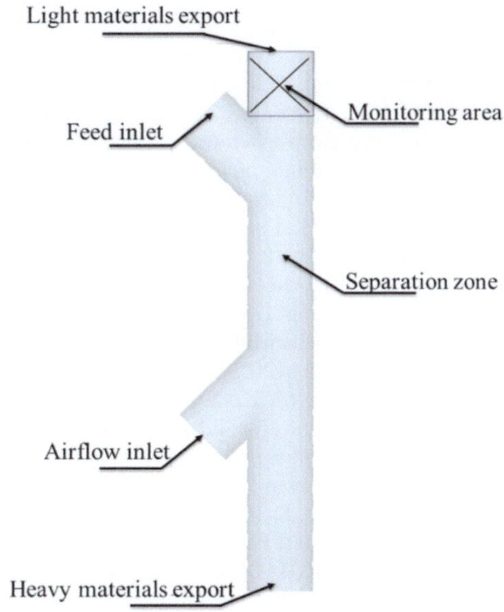

Fig. 17.12 Structure of the vertical pneumatic separator for lithium iron phosphate battery recycling [10]

includes a monitoring area to track particle behavior and evaluate separation efficiency, as well as two export outlets: one at the top for light materials and another at the bottom for heavy materials. This design leverages differences in particle density and aerodynamic properties to achieve effective separation, making it a cost-effective and scalable solution for industrial battery recycling.

The key findings highlight the importance of optimizing operational parameters for efficient separation. For instance, the airflow velocity significantly influences particle movement and recovery rates. The simulation results identify 2.0 m/s as the optimal velocity for separating plastic films and 3.4 m/s for anode copper and cathode aluminum foils. While higher velocities can accelerate particle movement, excessive speeds may introduce turbulence, reducing recovery rates. Similarly, the height of the separation zone plays a critical role. A height of 300–350 mm is ideal for plastic film separation, whereas 300 mm works best for metal foils. Deviations from these heights can lead to airflow concentration issues or unwanted eddy currents, compromising efficiency. Additionally, the initial velocity of particles affects separation time without significantly altering recovery rates. For metal foils, an initial velocity of 0.2 m/s is optimal to prevent particle buildup and ensure smooth operation.

References

1. J. Chen, M. Asachi, A. Hassanpour, M. Babaie, M. Jabbari, *Modelling of Lithium-Ion Battery Electrode Calendering: A Critical Review* (Elsevier, 2025). https://doi.org/10.1016/j.est.2025.116702

2. R. Ge, D.J. Cumming, R.M. Smith, Discrete element method (DEM) analysis of lithium ion battery electrode structures from X-ray tomography-the effect of calendering conditions. Powder Technol. **403**, 321 (2022). https://doi.org/10.1016/j.powtec.2022.117366
3. R. Ge, A.M. Boyce, Y.S. Zhang, P.R. Shearing, D.J. Cumming, R.M. Smith, Discrete element method and electrochemical modelling of lithium ion cathode structures characterised by X-ray computed tomography. Chem. Eng. J. **465**, 303–378 (2023). https://doi.org/10.1016/j.cej.2023.142749
4. A. Lundkvist, P.L. Larsson, E. Olsson, A discrete element analysis of the mechanical behaviour of a lithium-ion battery electrode active layer. Powder Technol. **425**, 16–89 (2023). https://doi.org/10.1016/j.powtec.2023.118574
5. Y.K. Lee, C. Sung, J. Kim, C. Hong, J. Choi, Multiphysics modeling of the influence of initial pressure on mechanical and electrochemical performance of all-solid-state batteries. J. Energy Stor. **82**, 202–221 (2024). https://doi.org/10.1016/j.est.2024.110431
6. F. Frankenberg, M. Kissel, C.F. Burmeister, M. Lippke, J. Janek, A. Kwade, Investigating the production of all-solid-state battery composite cathodes by numerical simulation of the stressing conditions in a high-intensity mixer. Powder Technol. **435**, 54–63 (2024). https://doi.org/10.1016/j.powtec.2024.119403
7. M. So, G. Inoue, K. Park, K. Nunoshita, S. Ishikawa, Y. Tsuge, Simulation of the compaction of an all-solid-state battery cathode with coated particles using the discrete element method. J. Power Sour. **530**, 112–145 (2022). https://doi.org/10.1016/j.jpowsour.2022.231279
8. M. So, S. Yano, A. Permatasari, T.D. Pham, K. Park, G. Inoue, Mechanism of silicon fragmentation in all-solid-state battery evaluated by discrete element method. J. Power Sour. **546**, 54–65 (2022). https://doi.org/10.1016/j.jpowsour.2022.231956
9. J. Huang, J. Hu, D. Zhang, Y. Du, C.Y. Wu, Q. Cai, Modelling heat conduction in 3D composite cathode microstructures of all-solid-state batteries. J. Energy Stor. **114**, 65–98 (2025). https://doi.org/10.1016/j.est.2025.115692
10. E. Dianyu et al., Numerical investigation and optimization of vertical pneumatic separation of film-like particles in lithium iron phosphate battery recycling. Sep. Purif. Technol. **361**, 654–856 (2025). https://doi.org/10.1016/j.seppur.2025.131551

Chapter 18
Machine Learning and Optimization

18.1 Machine Learning Strategies

Machine learning (ML) is a branch of artificial intelligence that enables computers to learn from data without being explicitly programmed. These techniques have gained prominence in many fields, including scientific computing, engineering and optimization, because of their ability to model complex systems and extract patterns from large datasets. In this section, the key principles behind machine learning strategies are introduced.

Supervised learning is one of the most widely used machine learning strategies. In supervised learning, the model is trained on a labeled dataset, where each input sample is paired with a corresponding output or target. The goal is for the algorithm to learn a mapping from inputs to outputs such that it can make predictions on unseen data. This strategy relies heavily on the principle of minimizing the error between the predicted output and the actual target. In the context of DEM, supervised learning can be applied to predict material behavior, particle interactions or system parameters based on previously observed simulations or experimental data. For example, a regression model could be used to predict the forces between particles in a granular flow, while classification models might predict the occurrence of certain phenomena, like the formation of clusters or the onset of failure in a particle system. In contrast to supervised learning, unsupervised learning does not require labeled data. The goal of unsupervised learning is to identify hidden patterns or structure in the data. Common methods in unsupervised learning include clustering, dimensionality reduction and anomaly detection. Clustering algorithms, like k-means or hierarchical clustering, group data points based on similarity, while dimensionality reduction methods like principal component analysis reduce the number of variables to uncover underlying structures in high-dimensional data.

Both supervised and unsupervised learning often employ a variety of models, among which artificial neural networks (ANNs) are computing systems inspired by

R. K. Agarwal and Y. Shao, *Theory and Application of Discrete Element Method to Simulation of Multiphase Flows*, https://doi.org/10.1007/978-3-032-11157-9_18

the biological neural networks in animal brains. An ANN is composed of interconnected layers of nodes also called "neurons". Each connection can transmit a signal to other neurons, and each neuron processes its inputs and sends an output to the next layer. The network "learns" by adjusting the weights of these connections to minimize the prediction error. A simple ANN with one or two hidden layers is often called a shallow network. When an ANN contains many hidden layers, it is considered a deep neural network (DNN), which falls under the subfield of deep learning. This depth allows them to learn increasingly complex, hierarchical features from the data. Several specialized architectures have been developed within deep learning:

(1) Deep neural networks

Deep neural networks (DNNs) are the most generic form, also known as multi-layer perceptrons. They consist of multiple fully-connected layers and are excellent for learning complex, non-linear relationships in high-dimensional data. For example, they can learn the mapping from DEM parameters to a bulk flow property.

(2) Convolutional neural networks

Convolutional neural networks (CNNs) are designed primarily for processing grid-like data, such as images. CNNs use mathematical "convolutions" instead of general matrix multiplication in at least one of their layers. They are exceptionally powerful for extracting spatial features. In DEM, while not dealing with images in the traditional sense, CNNs can be applied to voxelized particle data or 2D slices to identify patterns and structures in granular assemblies.

(3) Recurrent neural networks

Recurrent neural networks (RNNs) are specialized for sequential data, where the order of inputs matters (e.g., time series, language). RNNs have loops in them, allowing information to persist, effectively giving them a "memory" of previous inputs. Long short-term memory (LSTM) networks are a popular and powerful type of RNNs designed to learn long-term dependencies. In DEM, RNNs and LSTMs are suitable for predicting the temporal evolution of a system, such as forecasting particle velocities or stresses over time based on a sequence of past states.

Reinforcement learning (RL) is a paradigm distinct from supervised and unsupervised learning. In RL, an agent learns to make decisions by performing actions within an environment to maximize a cumulative reward. The agent is not told which actions to take but instead must discover which actions yield the most reward through trial and error. This paradigm is highly analogous to DEM parameter calibration: an RL agent could be tasked with adjusting parameters in a simulation, where the "reward" is inversely related to the error between the simulation output and experimental data. Through trial and error, the agent thus learns the sequence of parameter adjustments that leads to the most accurate calibration.

18.2 Research Focus of Applications of Machine Learning in DEM Simulations

The current research focus of applications of machine learning in DEM simulations are primarily concentrated on the following four interconnected directions, each contributing to smarter, faster and more reliable DEM-assisted engineering and science.

(1) Surrogate modeling for rapid prediction

A primary and highly active research focus is the development of highly accurate surrogate models based on a carefully generated dataset from a limited number of full-scale DEM simulations. These simulations may include particle properties, reactor parameters and operational parameters. Once trained, the surrogate model can predict key performance metrics, such as mass flow rate, mixing index or energy consumption, in milliseconds instead of the hours or days required for a full DEM run.

(2) Parameter calibration

Calibrating microscopic DEM parameters to match macroscopic experimental data is a classic, time-consuming challenge that often relies on manual trial and error. Machine learning is being used to automate and vastly accelerate this process. For example, an RL agent can be trained to intelligently adjust DEM parameters in a virtual environment, receiving a reward based on how closely the simulation output matches the target experimental data. This transforms calibration from a manual art into an efficient, automated and objective computational process, ensuring more reliable and reproducible simulation results.

(3) Analysis of complex phenomena and feature detection

Beyond mere parameter prediction, a more advanced research focus involves using ML for intelligent analysis and discovery within DEM data. DEM simulations generate vast, high-dimensional datasets containing the trajectory, force and velocity of every particle at every time step. ML techniques are being employed to mine this data for insights that are difficult for humans to discern. For instance, CNNs can be applied to voxelized snapshots of particle assemblies to automatically identify and characterize complex phenomena like shear band formation, particle mixing/segregation patterns or the onset of jamming. Similarly, RNNs and LSTMs are used to model the temporal evolution of these systems, forecasting long-term behavior or predicting failure modes from short-time historical data. This application moves ML from a simple predictor to an intelligent analytical tool for scientific discovery.

(4) Data-driven closure model

Another major research frontier involves developing data-driven closure relations to enhance the accuracy of computational models. It uses ML to derive precise constitutive laws and interaction models directly from high-fidelity data, avoiding reliance on idealized empirical correlations. For example, DNNs can be trained on DEM simulation data to learn complex stress–strain relationships in granular materials, forming constitutive models that bridge particle-scale physics with continuum-scale methods like FEM. Similarly, in coupled CFD-DEM simulations, key sub-models such as drag force, heat transfer, mass transfer and reaction kinetics, which are traditionally dependent on simplified formulas, can be replaced by ML-learned closure relations derived from detailed DNS-DEM data or experiments. These learned models serve as highly accurate surrogates that capture underlying physical interactions more faithfully than conventional formulas. By integrating such data-driven closures into multiscale simulations, this approach significantly improves predictive accuracy while maintaining computational efficiency, ultimately enabling more reliable modeling of complex granular and multiphase systems.

In summary, machine learning provides a powerful toolkit to overcome the computational bottlenecks of DEM, automate its setup, and extract deeper insights from its results. This synergy is advancing the field toward more efficient, accurate, and intelligent simulation of granular and particulate systems.

18.3 Example Applications of Machine Learning

Shao et al. [1] employed machine learning to analyze and predict flow patterns in a moving bed reactor for chemical looping combustion, using a dataset from prior CFD-DEM simulations. The mass flow index is used to characterize the flow, distinguishing between mass flow and funnel flow regimes. A random forest regression model first ranks the influence of various parameters on the MFI. It finds that the axial position has the dominant effect, followed by the wedge angle, reactor angle and the ratio of down-comer to reactor diameter. The solid mass flux has the least impact. Subsequently, a feed forward neural network is developed to predict solid axial velocity, using the key structural parameters and spatial positions as inputs. This feed forward neural network model is then used to map the relationship between structural parameters and the effective transition point height, which defines the boundary between flow patterns. Based on this analysis, the study recommends optimal reactor configurations to maximize the mass flow region, providing a data-driven tool for reactor design and optimization without the need for further complex simulations.

Kalay et al. [2] developed an ANN model to rapidly predict the mass flow rate in a screw conveyor, as shown in Fig. 18.1. The ANN is trained on data generated from a limited set of DEM simulations. The model takes into account both screw conveyor parameters (rotational velocity, inclination angle) and granular material properties (particle density, mean diameter, sliding and rolling friction coefficients, coefficient of restitution). The model shows excellent agreement for horizontal conveyor configurations, with errors generally below 3%.

Lu et al. [3] presented a study on accelerating DEM simulations of granular flows using a CNN. The core innovation lies in replacing the most expensive component of DEM, that is the explicit calculation of collision forces between particles and walls,

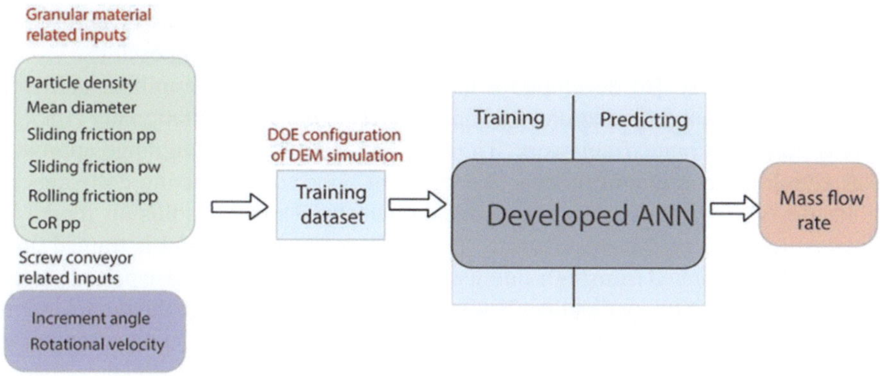

Fig. 18.1 Mass flow rate prediction of screw conveyor using artificial neural network method [2]

with a predictive model based on a CNN. The high-fidelity training data is generated from traditional MFiX-DEM simulations of granular flows in two benchmark systems, a rotating drum and a hopper. A physics-inspired multi-scale loss function is developed, which combines a micro-scale loss, which ensures accuracy at the individual particle level, with a macro-scale loss, which ensures the correct prediction of bulk properties like the system's center of mass. The model achieves a 78 × speed-up over the conventional DEM solver. This performance gain is primarily enabled by the CNN's ability to handle time steps 500 times larger than those required for numerical stability in a physics-based collision resolution.

Further, Xu and Shen [4] presented an improved ML approach for efficiently predicting granular flows. The novel ML approach is physically consistent with Newtonian mechanics. A continuous CNN is developed to first predict the instantaneous particle acceleration resulting from collisions. Crucially, the network architecture processes particle–particle and particle–wall interactions in two separate paths, allowing for dedicated feature learning for each collision type. The predicted acceleration is then used to update the particle velocity and, subsequently, the position via integration, mirroring the fundamental DEM calculation sequence.

Cheng and Lin [5] used a ML based on a DNN to predict the formation of force chains in dense granular flows, a critical mesoscale phenomenon that governs macroscopic behaviors like non-local effects and jamming. The model is trained on DEM data from inclined plane flows, using 14 input features per particle, including novel topological parameters like local clustering coefficient, to perform binary classification. It achieved high accuracy of 94.7% and perfect recall of 100%, especially in non-Bagnold flows where force chains dominate. Analysis via Shapley Additive exPlanations (SHAP) values reveals that the coordination number and the local clustering coefficient are the two most important features for the prediction, aligning with the physical understanding that force chains form in densely connected, clustered regions. Additionally, a reduced set of seven experimentally measurable features is defined. When using this practical feature set, the model maintained reasonable performance, demonstrating its potential for inferring force chains in real-world granular flow experiments where direct contact measurement is impossible.

Adamczyk et al. [6] developed a machine learning-based simplified collision model to enhance the simulation of granular flows within a computationally efficient hybrid Euler–Lagrange framework. The standard hybrid Euler–Lagrange approach uses simplified continuum models that perform poorly in predicting collisions in dilute flows. To address this, the conventional physics-based collision model is replaced with a data-driven surrogate. The machine learning model is trained on a large dataset generated from high-fidelity DEM simulations of pairwise particle collisions. Using the eXtreme Gradient Boosting (XGBoost) algorithm, the model learns the complex relationship between pre-collision particle states and their post-collision velocities. This allowed the hybrid model to accurately resolve individual collisions in a Lagrangian manner, much like DEM, but without its prohibitive computational cost. Validation against experimental data shows that the model successfully predicts particle velocity profiles and captures the dispersion of colliding particle streams, a phenomenon the standard hybrid Euler–Lagrange model fails to reproduce.

Niu et al. [7] presented a hybrid, physically informed framework for calibrating DEM parameters for cohesive wet coal particles across a range of moisture contents (5–25%) to address the challenge of accurately simulating moisture-dependent adhesion and friction behaviors. The methodology integrates response surface methodology, artificial neural networks, and a multi-phase genetic algorithm. Initially, experimental data are collected through contact mechanics tests and repose angle measurements. Response surface methodology is used to establish preliminary relationships between DEM parameters and the repose angle. Artificial neural network is then employed to model the highly nonlinear interactions that response surface methodology can not fully capture, especially under high-moisture conditions. Finally, multi-phase genetic algorithm is applied to optimize the parameters in three phases: initial global search, integration of physical constraints and targeted refinement under high-moisture conditions. The optimized DEM parameters, including surface energy, rolling friction coefficients and restitution coefficients, yield repose angle predictions with relative errors consistently below 2%.

Westbrink et al. [8] presented a method for calibrating DEM using multi-objective ML approach. The method involves an Actor-Critic RL agent that interacts with two distinct DEM simulation environments: one for the static angle of repose measured via a lifting cylinder test and another for the dynamic angle of repose measured via a rotating drum test. The agent's state is defined by four key DEM parameters including coefficient of friction, coefficient of restitution, coefficient of rolling friction and a cohesion parameter, and its actions involve adjusting these parameters. The reward function is a weighted sum of rewards from both environments, calculated based on how closely the simulated angles of repose match the target values obtained from physical experiments on a mobile calibration unit. The method was successfully validated on three materials including plastic granulate, wood pellets and wet sand, accurately matching experimental angle-of-repose targets and performing well in a separate validation test.

18.4 Bayesian Optimization and Genetic Algorithms

Optimization is a critical component of many computational methods. It refers to the process of finding the best solution to a problem, typically by minimizing or maximizing an objective function subject to certain constraints. In the context of DEM simulations and other complex systems, optimization can be used to fine-tune parameters, enhance system performance and achieve desired outcomes with greater efficiency. This section discusses two popular optimization strategies: Bayesian optimization and genetic algorithms (GAs), exploring their principles, applications and how they are employed to solve DEM problems.

Bayesian optimization is a probabilistic model-based optimization technique, particularly suitable for high-dimensional, noisy and expensive-to-evaluate functions. It treats the function as a black box and uses a probabilistic surrogate model, typically a Gaussian process, to approximate the objective. This model provides both predicted values and uncertainty estimates across the parameter space. An acquisition function then guides the selection of subsequent sample points by balancing exploration (uncertain regions) and exploitation (promising regions). In DEM simulations, where each evaluation can be resource-intensive, Bayesian optimization reduces the number of simulations required to identify optimal parameters, such as particle size distributions, contact properties or material stiffness, thereby accelerating calibration and improving simulation accuracy.

GAs are a class of evolutionary algorithms inspired by the process of natural selection. They are particularly effective in solving optimization problems where the solution space is large, nonlinear or poorly understood. The process begins with the creation of an initial population of candidate solutions. Each solution is typically represented as a string of parameters or genes, which encode potential solutions to the problem. These candidates are evaluated based on a fitness function that measures how well each solution solves the problem. The best-performing individuals are then selected to "reproduce", with crossover and mutation used to create new offspring. These offspring then replace the less successful solutions, and the process repeats for a predefined number of generations. GAs are highly flexible and can be applied to a wide range of optimization problems, including those involving complex, nonlinear objective functions. In DEM, genetic algorithms can optimize diverse parameters including particle packing configurations, contact models and boundary conditions. They are especially valuable when aiming to achieve global optima in systems with multiple local minima, such as granular flow dynamics or cohesive particle interactions.

While both methods are powerful, they suit different optimization scenarios. Bayesian optimization excels when function evaluations are expensive and the parameter space is of moderate dimensionality. It efficiently minimizes the number of evaluations through intelligent sampling, but may struggle in very high-dimensional settings. GAs, on the other hand, are robust and parallelizable, making them ideal for complex and highly nonlinear problems. They do not require differentiable objective functions and can explore vast parameter spaces, though often at the cost of higher computational effort. In DEM, the choice between these methods depends on factors such as computational budget, parameter dimensionality, and the nature of the simulation output.

18.5 Example Applications of Optimization Algorithms

Hadjidoukas et al. [9] introduced a Bayesian framework for uncertainty quantification and propagation in DEM simulations of granular materials. It calibrates three force–displacement contact models using experimental data from steel particle–wall collisions, employing a parallel Transitional Markov Chain Monte Carlo method to estimate parametric uncertainties and perform model selection. Results show that both parametric and model errors are significant for robust predictions. The calibrated uncertainties are propagated to two application cases: the "Brazil Nut" segregation effect and silo discharge. In both cases, output quantities like segregation time and avalanche size exhibit substantial uncertainty, strongly influenced by contact model parameters.

Huang et al. [10] developed a computational shape optimization method to redesign hoppers for improved granular flow. Combining a FEM continuum model with a hybrid GA and gradient descent method, the approach automatically evolves an initial conical shape into an optimized curved profile. The objective is to maximize the mass flow zone, the region where particle velocity exceeds a defined threshold. The optimized designs increase the mass flow zone fraction by over 90%. Sensitivity analysis shows the method is most effective for hoppers with initial half-angles of 38°–45° and for materials with high internal friction.

Do et al. [11] proposed an automated calibration framework for DEM parameters using a multi-objective GA to address the challenge of selecting accurate microscopic input parameters that reproduce the macroscopic behavior of granular materials. The proposed methodology consists of three main phases. First, a database is generated by running DEM simulations across a multi-dimensional grid of sampled input parameter values to capture a range of macroscopic responses. Second, the Non-dominated Sorting Genetic Algorithm II (NSGA-II) is employed to solve the calibration as a multi-objective optimization problem. This approach simultaneously minimizes both the simulation error and the simulation time. The use of NSGA-II allows for the visualization of trade-offs between these conflicting objectives via a Pareto-optimal front, providing a set of optimal solutions rather than a single point. In the third phase, the calibrated parameters are verified and validated through additional DEM simulations and comparison with experimental data.

Mohajeri et al. [12] presented a calibration framework for DEM simulations of cohesive bulk materials. NSGA-II is used to determine optimal DEM parameters. The algorithm minimizes the difference between simulated and experimental shear stresses under two pre-consolidation stress levels. Three parameters are calibrated: inter-particle sliding friction coefficient, surface adhesion energy and constant pull-off force. The calibration is performed efficiently using a reduced simulation domain with cylindrical periodic boundaries, which cuts computational time in half while maintaining accuracy.

Shafaie and Rad [13] developed an automated calibration framework for DEM simulations of colored self-compacting concrete using a multi-objective GA to optimize six micro-parameters of a linear parallel bond model based on experimental uniaxial compressive strength test results. The calibration process is fully automated: GA in MATLAB drives DEM simulations in PFC3D, evaluates the results using a fitness function derived from mean squared error, and iteratively adjusts micro-parameters until the numerical outputs closely match experimental values. Fitness values consistently exceed 0.94, indicating high calibration accuracy.

References

1. Y. Shao, R.K. Agarwal, X. Wang, B. Jin, Study of flow patterns in a moving bed reactor for chemical looping combustion based on machine learning methods. J. Energy Resour. Technol. **145**(6), 6562 (2023). https://doi.org/10.1115/1.4056562
2. E. Kalay, M.E. Boğoçlu, B. Bolat, Mass flow rate prediction of screw conveyor using artificial neural network method. Powder Technol. **408**, 659 (2022). https://doi.org/10.1016/j.powtec.2022.117757
3. L. Lu, X. Gao, J.F. Dietiker, M. Shahnam, W.A. Rogers, Machine learning accelerated discrete element modeling of granular flows. Chem. Eng. Sci. **245**, 6548 (2021). https://doi.org/10.1016/j.ces.2021.116832
4. D. Xu, Y. Shen, An improved machine learning approach for predicting granular flows. Chem. Eng. J. **450**, 78 (2022). https://doi.org/10.1016/j.cej.2022.138036
5. C.H. Cheng, C.C. Lin, Prediction of force chains for dense granular flows using machine learning approach. Phys. Fluids **36**(8), 321 (2024). https://doi.org/10.1063/5.0216408
6. W. Adamczyk et al., A machine learning-based simplified collision model for granular flows. Powder Technol. **444**, 654 (2024). https://doi.org/10.1016/j.powtec.2024.120006
7. F. Niu et al., Physically informed machine learning-enhanced DEM parameter calibration for cohesive wet coal. Powder Technol. **467**, 1523 (2026). https://doi.org/10.1016/j.powtec.2025.121523
8. F. Westbrink, A. Elbel, A. Schwung, S.X. Ding, Optimization of DEM parameters using multi-objective reinforcement learning. Powder Technol. **379**, 602–616 (2021). https://doi.org/10.1016/j.powtec.2020.10.067
9. P.E. Hadjidoukas, P. Angelikopoulos, D. Rossinelli, D. Alexeev, C. Papadimitriou, P. Koumoutsakos, Bayesian uncertainty quantification and propagation for discrete element simulations of granular materials. Comput. Methods Appl. Mech. Eng. **282**, 218–238 (2014). https://doi.org/10.1016/j.cma.2014.07.017
10. X. Huang, Q. Zheng, D. Liu, A. Yu, W. Yan, A design method of hopper shape optimization with improved mass flow pattern and reduced particle segregation. Chem. Eng. Sci. **253**, 579 (2022). https://doi.org/10.1016/j.ces.2022.117579
11. H.Q. Do, A.M. Aragón, D.L. Schott, A calibration framework for discrete element model parameters using genetic algorithms. Adv. Powder Technol. **29**(6), 1393–1403 (2018). https://doi.org/10.1016/j.apt.2018.03.001
12. M.J. Mohajeri, H.Q. Do, D.L. Schott, DEM calibration of cohesive material in the ring shear test by applying a genetic algorithm framework. Adv. Powder Technol. **31**(5), 1838–1850 (2020). https://doi.org/10.1016/j.apt.2020.02.019
13. V. Shafaie, M.M. Rad, Multi-objective genetic algorithm calibration of colored self-compacting concrete using DEM: an integrated parallel approach. Sci. Rep. **14**(1), 715 (2024). https://doi.org/10.1038/s41598-024-54715-4

Index

R. K. Agarwal and Y. Shao, *Theory and Application of Discrete Element Method to Simulation of Multiphase Flows*, https://doi.org/10.1007/978-3-032-11157-9

MIX
Papier aus verantwortungsvollen Quellen
Paper from responsible sources
FSC® C105338

If you have any concerns about our products,
you can contact us on
ProductSafety@springernature.com

In case Publisher is established outside the EU,
the EU authorized representative is:
Springer Nature Customer Service Center GmbH
Europaplatz 3, 69115 Heidelberg, Germany

Printed by Libri Plureos GmbH
in Hamburg, Germany